Water Droplets to Nanotechnology
A Journey Through Self-Assembly

Water Droplets to Nanotechnology
A Journey Through Self-Assembly

Plinio Innocenzi and Luca Malfatti
Laboratorio di Scienza dei Materiali e Nanotecnologie (LMNT), Piazza Duomo, Italy
Email: plinio@uniss.it

Paolo Falcaro
CSIRO, Victoria, Australia
Email: paolo.falcaro@csiro.au

RSC Publishing

ISBN: 978-1-84973-664-0

A catalogue record for this book is available from the British Library

Published by The Royal Society of Chemistry,
Thomas Graham House, Science Park, Milton Road,
Cambridge CB4 0WF, UK

Registered Charity Number 207890

Visit our website at www.rsc.org/books

Printed in the United Kingdom by Henry Ling Limited, Dorchester, DT1 1HD, UK

Preface

THINK BIG AND...START NANO...

Within the nano-world, self-assembly is the result of a delicate balance among different chemical-physical forces, and one of the most striking properties of the nano-objects is the ability to self-organize into complex structures. Following the consideration that we are all the result of a self-organization process, we should be aware that the complexity of living structures holds the secret to self-assembly.

The self-organization depends on different variables such as composition, shape and dimension of the involved objects, as well as their different physical states.

Liquid systems have exceptional features that make them excellent candidates for self-assembly.

Firstly, liquids offer the possibility to dissolve or suspend different species such as copolymers, nanoparticles, biomolecules and biological entities. Secondly, chemical reactions occur homogenously and quickly in the liquid state. Finally, the liquid can be removed, taking advantage of the evaporation process.

Despite evaporation being a common phenomenon in everyday life, it plays a crucial role in the self-organization of systems composed of liquids and nano-objects (colloidal systems). These are either pre-formed solids such as nanoparticles, nanorods of sheets, or supramolecular structures formed during the self-assembly process itself. That is the case for surfactants, which are a fascinating class of molecule. Under solvent evaporation surfactants are able to generate a large variety of super-structures from basic to complex shapes.

However, there is a drawback in this business and that is the instability of the process. When evaporation is involved, most of the

Water Droplets to Nanotechnology: A Journey Through Self-Assembly
By Plinio Innocenzi, Luca Malfatti and Paolo Falcaro
© P. Innocenzi, L. Malfatti and P. Falcaro 2013
Published by the Royal Society of Chemistry, www.rsc.org

self-assembly processes are not equilibrium phenomena because they are kinetically controlled.

Achieving organization through non equilibrium sounds like an ambitious task but fortunately it works!

In this book, we have highlighted that evaporation by itself does not create organization; order is indeed driven by the forces that arise during evaporation. These forces can trigger the system to become ordered or disordered. In the latter case, the final material will not exhibit any organization at the nanoscale; only a careful control of the evaporation rate allows dancing on the tightrope of self-assembly.

This book can be treated as a small journey that reveals the secrets of self-assembly occurring during evaporation of a colloidal water droplet; little by little, the journey moves from the mysteries of a coffee stain on glassware to the formation of complex hierarchical systems for advanced multifunctional applications.

READING THIS BOOK

Nowadays, expertise in a specific field is mandatory for developing cutting edge science. However, this requirement can lead to a loss of the wider perspective, which is the basis of a multidisciplinary approach.

This book has been written with the aim of providing a general overview of self-organization processes whose complexity and potential are difficult to understand without some basic knowledge. Therefore, this book should not be considered as a comprehensive scientific treatise but rather as a journey into nanotechnology following the path of self-assembly. To keep the focus on this intriguing phenomenon, each chapter proposes a selected number of techniques and results. For example, a variety of fabrication techniques allows one to obtain specific optical devices called *photonic crystals* and Chapter **6** highlights cases involving both evaporation and self-assembly. Not all the literature in the field has been cited; to keep the story on the right path we have carefully selected the references based on our personal opinion.

The presented journey should be enjoyable to the reader and the complexity of the different topics has been simplified. Mathematical formalisms have been avoided unless strictly necessary and we have tried to explain concepts with figures and schematics where relevant.

In summary, this book is not aimed at scientists with a specific background in self-assembly, but at curious readers interested in gaining a general understanding of the topic. And finally, Professor

David Evans of the Beijing University of Chemical Technology is gratefully acknowledged for critical reading of this book and his highly appreciated comments.

Plinio Innocenzi
Luca Malfatti
Paolo Falcaro

Contents

Water Droplets to Nanotechnology: A Journey Through Self-Assembly
By Plinio Innocenzi, Luca Malfatti and Paolo Falcaro
© P. Innocenzi, L. Malfatti and P. Falcaro 2013
Published by the Royal Society of Chemistry, www.rsc.org

Author biographies

Plinio Innocenzi is a Professor of Materials Science at the University of Sassari, Italy and Director of the Laboratory of Materials Science and Nanotechnology (LMNT). He graduated in physics from the University of Padua, Italy and was an Assistant Professor at the Department of Mechanical Engineering, Material Section at the same university. In 1994, he was awarded a Science and Technology Fellowship in Japan from the European Commission and became an associate foreign researcher at Kyoto University, Japan. He has also worked as a visiting Professor at the University Paris VI, Kyoto University, Osaka Prefecture University and Beijing University of Chemical Technology. He was one of the founders of the International Sol–Gel Society and has served on the Board of Directors. In addition, he is a Fellow of the Royal Society of Chemistry. He has authored more than 150 articles in international ISI journals, 3 books and 12 patents. His research interests are focused on self-assembly at the nanoscale, sol–gel chemistry and hybrid organic–inorganic materials.

Water Droplets to Nanotechnology: A Journey Through Self-Assembly
By Plinio Innocenzi, Luca Malfatti and Paolo Falcaro
© P. Innocenzi, L. Malfatti and P. Falcaro 2013
Published by the Royal Society of Chemistry, www.rsc.org

Luca Malfatti (born 1977 in Venice, Italy) is an Assistant Professor of Materials Science at the University of Sassari, Italy. He received a Master's degree in Material Science from the University of Padua in 2004 and a Ph.D. degree in "*Materials for environment and energy*" from the University of Rome Tor Vergata in 2010. He has worked as a visiting researcher at the University Paris VI, the CNEA in Buenos Aires, Osaka Prefecture University and the CSIRO Materials Science and Engineering Division in Melbourne. He has authored more than 70 articles in international ISI journals. His research deals with the synthesis of mesoporous and hierarchical porous materials obtained by supramolecular self-assembly and their application as advanced functional ceramics.

Paolo Falcaro is a materials scientist who received his Ph.D. in Material Engineering in 2006 from Bologna University. From 2005 to 2008, he worked as a research scientist in the Nanofabrication Facility (Civen/Nanofab, Venice, Italy) for industrial applications using sol–gel technologies. In 2009, he joined CSIRO (Material Science and Engineering Division, CMSE in Melbourne, Australia) as a Postdoctoral Fellow. He is currently a research scientist at CSIRO and an ARC Discovery Early Career Research Fellow. He is the recipient of several national and international awards, including the Ulrich Award, the Japan–Australia Emerging Leader Award and the CSIRO Julius Award. He investigates self-assembled porous materials preparation and patterning and fabrication techniques to control the formation of functional materials. He also works on functional nanoparticles for sensing applications and biolabelling.

CHAPTER 1

The Coffee Stain: Using a Water Droplet for Self-assembly

This first chapter is dedicated to showing how a simple phenomenon, such as the evaporation of a droplet of a colloidal solution, can become a sophisticated tool for self-assembly and the fabrication of nanodevices. The chapter will be also be used to explain what the coffee-stain effect is and how it is generally understood and applied for designing patterns that self-organize upon droplet evaporation.

1.1 THE COFFEE-STAIN EFFECT

Drinking a cup of coffee, especially a good Italian espresso, is one of those little pleasures of everyday life. It can be difficult to imagine, therefore, a small droplet of our coffee as the starting point of a journey into self-assembly and nanotechnology as we try to enjoy our moments of relaxation. However, there is so much interesting and unexpected physics and chemistry behind the evaporation of a coffee droplet that it is worth observing the details of this phenomenon. The coffee-stain effect is intriguing because after the evaporation, what is left behind is not a homogeneous halo. It is a ring (**Figure 1.1**). This is a general effect as it does not matter what kind of solution is used as the coffee stain seems to be an ubiquitous phenomenon. The effect can be reproduced using different solvent−solute combinations, which comprise single molecules[1], nanoparticles[2], polymers[3], salts[4] and even bacteria; it is also observed at different length scales, from micron-sized particles to nano-objects as well as on the molecular level. It can also be observed on different types of substrates, such as silicon, glass, metal, mica and concrete. The coffee stain effect may represent an issue if homogeneous

Water Droplets to Nanotechnology: A Journey Through Self-Assembly
By Plinio Innocenzi, Luca Malfatti and Paolo Falcaro
© P. Innocenzi, L. Malfatti and P. Falcaro 2013
Published by the Royal Society of Chemistry, www.rsc.org

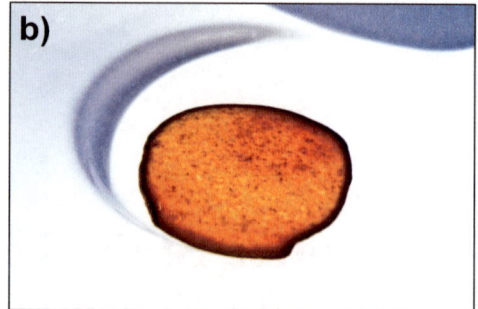

Figure 1.1   Dropping a coffee droplet on a solid substrate leaves upon evaporation a typical stain with a ring b) Enlarged image of the coffee stain presenting the typical ring shape (pictures by Laura Villanova).

spots are required. For instance, without preventing such an effect, the ink-jet technology would produce dishomogeneous plots[5,6]. A detailed understanding of the formation mechanism is necessary either to prevent the formation of the coffee-stain effect, or to take advantage of it in the development of new applications.

The first comprehensive explanation was reported by Deegan and co-workers in an article published in *Nature* (1997) and it is based on the observation that a mechanical constraint of an evaporating droplet causes a capillary flow towards the contact line[7]. When we pour a liquid droplet and observe how its evaporation occurs two different scenarios can be envisaged: in the first case the droplet is free to move on the substrate and in the second case the surface irregularities cause a blockage of the contact line that remains pinned until the evaporation process has gone almost to completion.

In the first case, therefore, the contact line is not pinned and the droplet shrinks; the profile of the droplet will not change while its radius, R, decreases (**Figure 1.2**). In contrast if the contact line is pinned[8] the droplet will flatten modifying its profile while the liquid evaporating from the edge is replenished by the liquid from the interior (**Figure 1.3**). If the liquid contains dispersed particles they will be carried out to the edge of the droplet by capillary flow that is induced by the pinning. The mass accumulation at the contact line then forms a ring-like structure around the droplet.

We can have a closer look at the evaporation process as described by Deegan[9] for an ideal case of a small, thin, dilute and circular drop with a fixed radius, R. The contact line is pinned and at every point of the droplet, r, and there is an evaporative flux, $J(r)$, which reduces the height, $h(r)$, of the droplet. The evaporation rate, however, is not

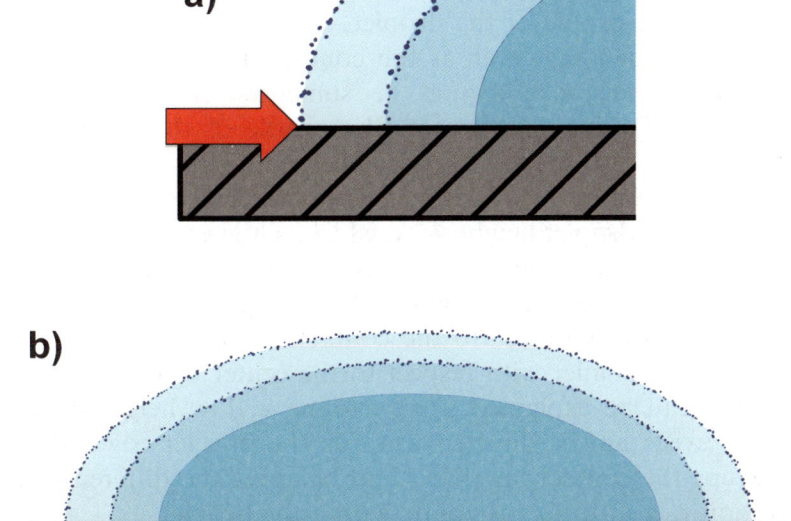

Figure 1.2 Evaporation without flow. (a) The contact line is not pinned and moves in the direction indicated by the red arrow; the droplet shrinks. (b) Evaporation occurs over the entire drop surface and if the contact line is free to recede, the drop profile is preserved during evaporation.

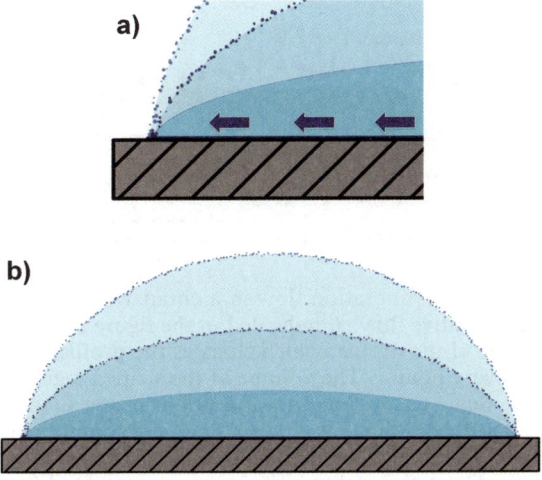

Figure 1.3 Evaporation without flow. (a) The contact line is pinned and the droplet does not shrink. A compensating flow (in the direction of the blue arrows) is needed to keep the contact line fixed; (b) the drop profile is not preserved during evaporation.

uniform and is greater when close to the edge of the contact line. If the contact line is not pinned, what is observed is a decrease of the radius while the overall profile of the droplet, as we have just seen, remains constant with time. However, if the contact line is pinned, the high evaporation at the edge should be somehow compensated by an outward flow. During this process the profile of the droplet should maintain its spherical shape, which is governed by the surface tension of the liquid. In a unit of time, Δt, the volume, which is removed, produces a decrease in the droplet height, $h(r)$, which is higher toward the center. If, however, we look to the flux it is larger at the edge (red arrows in the **Figure 1.4**) and in this case it would be expected that the volume of liquid that is removed close to the contact line during evaporation is greater and not smaller with respect to the center of the droplet. This observation can be explained if we assume a flow height of average speed $\vec{v}(r)$ exists and is directed towards the droplet edge. This flow height keeps the contact line fixed and balances the differences in the evaporation rate at different points within the droplet.

Why is the evaporative flow at the edge of the contact line enhanced? This is a complex problem which is similar to the case of an electric field near the sharp edge of a conductor. When close to the contact line, the evaporative flow diverges and is dependent on the relative distance from the center of the droplet and the contact angle[7,9]:

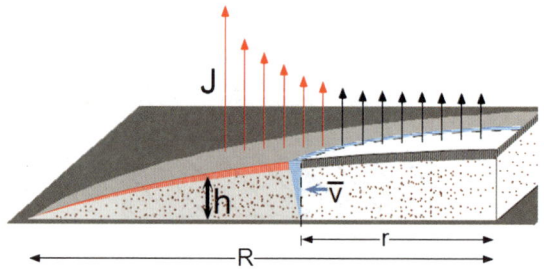

Figure 1.4   The outward evaporation flow in a circular pinned droplet of radius, R. The evaporative flow, J, indicated in the figure by the arrows, produces a shrinkage of the droplet which changes the profile and reduces its height, $h(r)$ at every point r. The volume of the striped region corresponds to the volume removed by J with evaporation; in the shaded region, however, the red-striped volume is smaller than the volume removed by the local flux J (red arrows), which is higher with respect to the center of the droplet. The deficit volume has to be compensated by a liquid flow: in a time Δt the fluid at r sweeps out the blue-striped region; its volume is the deficit volume and $\vec{v}(r)$ is its depth-averaged speed. (Reprinted by permission from Macmillan Publishers Ltd: *Nature*, 1997, **389**, 827, copyright 1997.)

$$J(r, t) \sim (R - r)^{-\lambda} \qquad (1.1)$$

where $\lambda = (\pi - 2\theta_c)/(2\pi - 2\theta_c)$ and θ_c is the contact angle of the liquid with the solid substrate. The reason for the higher evaporation rate at the edge is because of the greater probability of escaping, which a molecule that is close to the contact line has than a molecule at the center of the droplet. As we can see in **Figure 1.5**, the curvature at the edge decreases the probability of an evaporating molecule to be reabsorbed after evaporation. The same random path of an evaporating molecule at the center or at the edge of the droplet gives two different results: reabsorbing or escaping.

It should be noted that while pinning the droplet on the substrate at the beginning is activated by surface roughness and irregularities, a 'self-pinning' effect is also intrinsic to ring formation[8]. The accumulation of matter at the droplet edge increases the energy barrier that the contact line has to overcome in order to move.

After the work of Deegan, the physics of the coffee-stain phenomenon has been widely investigated and several complex models, which consider the influence of different parameters have been described. Further reading on the subject can be found in references[10–15].

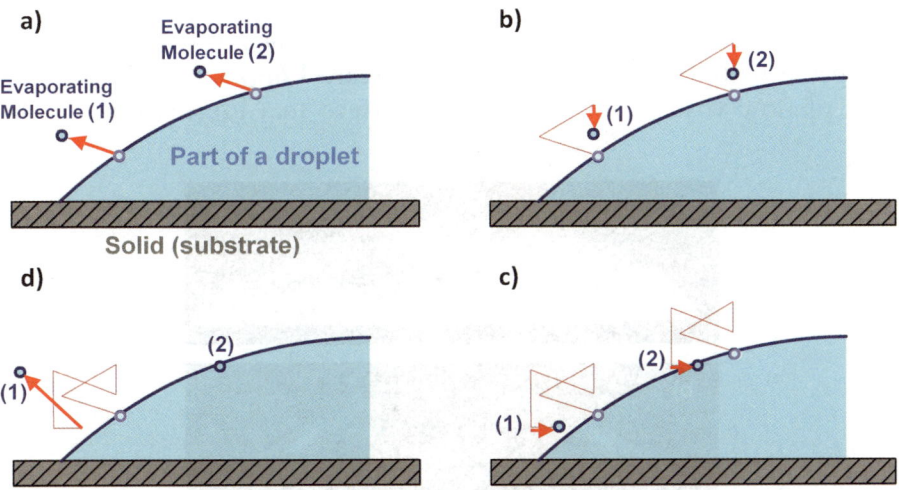

Figure 1.5 The probability of an evaporating molecule to be reabsorbed after evaporation depends on its position at the liquid−air interface. The same random path of an evaporating molecule at the edge (molecule 1) or at the center (molecule 2) of the droplet (a and b) gives two different results, reabsorbing or escaping (c and d).

1.2   REVERSING THE COFFEE STAIN

The model of Deegan[7] is based on an aqueous system and the assumption that for water the Marangoni effect is weak. Is this always true for other liquids? We will find out in the next chapter that the Marangoni effect can be used for self-assembly. However, for clarity, we can just briefly remind ourselves what this phenomenon is. The coffee stain is an example of flow induced by capillary forces but a flow in a liquid can be generated by a gradient of surface tension. In this case the liquid will flow from a low to a high surface tension region and this is known as the Marangoni effect. This difference in surface tension can arise either by concentration or temperature gradients. During an evaporation process thermal gradients can form easily, such as in the case of an evaporating droplet, and Marangoni flow cannot be discharged. The gradients can disrupt the internal liquid flow of the surface droplets with the formation of vortexes that interfere with the accumulation of material at the edge. In some simple organic liquids it has been observed experimentally that surface tension gradients, which are due to latent evaporation heat, cause a recirculatory flow and the deposition of colloidal particles occurs preferentially at the center rather than at the droplet edge[16]. The experimental evidence of the Marangoni-reversed coffee-stain phenomenon has been obtained by recording the flow field in a drying octane droplet containing 4.7 μm poly(methyl methacrylate) (PMMA) monodispersed fluorescent particles. These particles have been used to track the flow in the octane droplet while the experimental data have been compared with theoretical simulations. **Figure 1.6(a)** shows the appearance of flow vortexes on the droplet surface that are induced by the

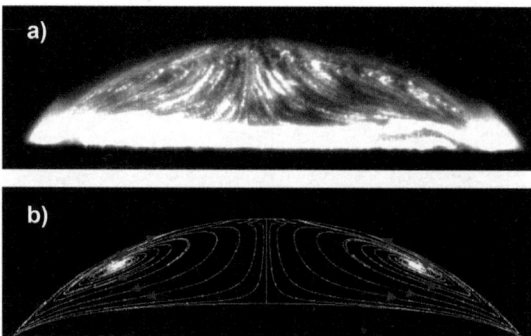

Figure 1.6   The Marangoni vortexes appear from the flow field in a drying octane droplet, (a) imaged experimentally and (b) predicted from modelling the evaporation process. (Reprinted with permission from H. Hu *et al.*, *Langmuir*, 2005, **21**, 3963. Copyright 2005 American Chemical Society.)

Marangoni effect. The bright region near the substrate is generated by a lens effect. The Marangoni stress on the microflow in an evaporating droplet[17] is, therefore, an important effect that is able to reverse coffee ring formation.

The Marangoni vortexes that form in the evaporating droplet are responsible for the recirculation of the particles; they are pushed from the surface towards the inner part of the droplet where they are adsorbed onto the substrate or eventually carried back towards the edge; reflowing to the top of the droplet. The theoretical streamlines of the Brownian dynamics simulations, which have been elaborated upon, take into account the heat transfer in a slowly evaporating droplet with a pinned contact line (**Figure 1.6(b)**) and are in agreement with experimental observations and the formation of Marangoni vortexes. The results mean that the conditions for observing coffee stains not only require a pinning of the contact line and a high evaporation rate at the droplet edge, but also a suppression of the Marangoni effect, which is caused by the latent heat of evaporation.

One question remains; why did Deegan not observe the Marangoni effect in his experiments? It was not an experimental error, several other experiments have, in fact, confirmed the results, but simply it has been found that for aqueous systems the Marangoni effect is very weak. The measured (right half in **Figure 1.6**) and simulated (left half) streamlines in a drying water droplet confirm that Marangoni flows are very weak in water, most likely due to contaminants that are always difficult to remove in aqueous systems and reduce the surface tension on the droplet surface[17].

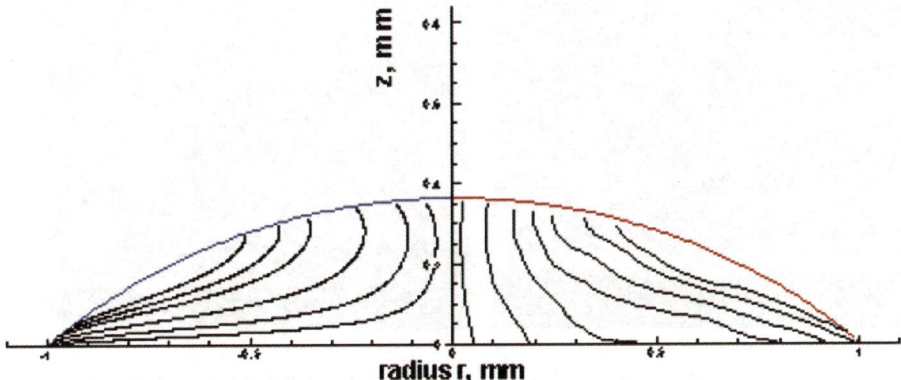

Figure 1.7 Streamlines in a drying water droplet, as measured (right half of droplet) and predicted (left half). (Reprinted with permission from H. Hu *et al., J. Phys. Chem. B,* 2006, **110**, 7090. Copyright 2006 American Chemical Society.)

1.3 SELF-ASSEMBLY PARTICLE RINGS

It should be quite clear that the coffee-stain effect, beside its negative implications in ink-jet printing, offers a unique opportunity of creating regular structures *via* self-assembly using particles of different dimensions, compositions and even shapes. The flow that accumulates the matter at the droplet edge can form regular rings using particles from micro to nanosizes with a large variety of functional properties. An example is the self-organization of silver nanoparticle rings[18]; the evaporation of water droplets containing 20–30 nm silver nanoparticles leaves very dense metallic rings behind at the edge. The close packing of the nanoparticles in the ring forms a dense structure whose electrical conductivity is higher (up to 15%) with respect to that of bulk silver even without any post annealing process. The optical (**Figure 1.8(a)**) and scanning electron microscope (SEM) (**Figure 1.8(b)**) images of a silver nanoparticle ring are

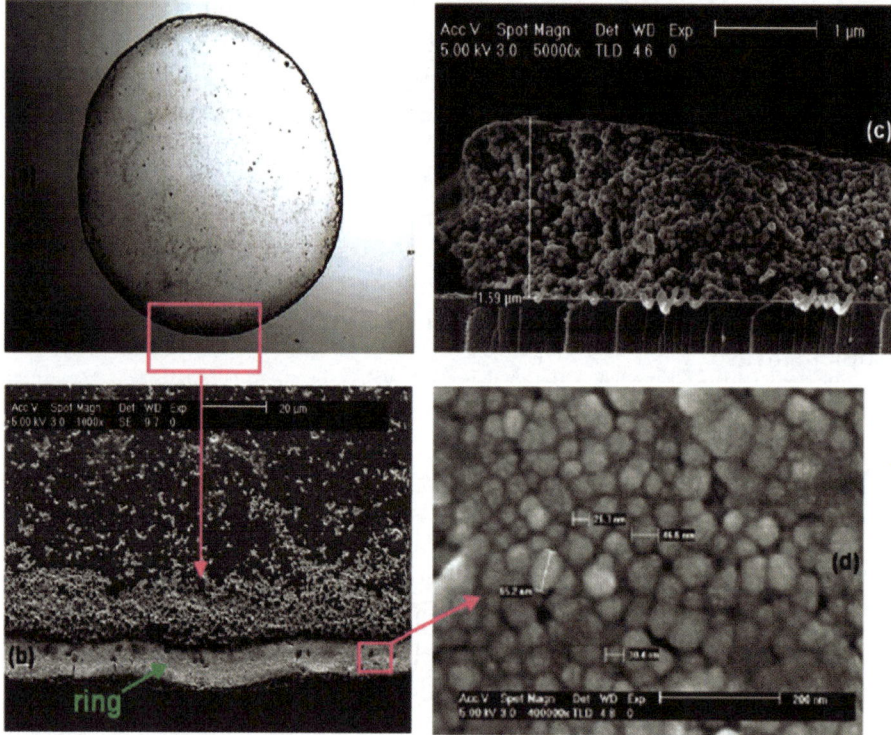

Figure 1.8   (a) Optical image of a 2 mm silver ring formed on a glass substrate; (b) SEM top-view detail of the same ring which is formed by an outer region of well packed silver nanoparticles and an inner rim of more loosely bounded particles; (c) and (d) are cross-sectional and top SEM images of the ring, respectively. (Reprinted with permission from S. Magdassi *et al.*, *Langmuir*, 2005, **21**, 10264. Copyright 2005 American Chemical Society.)

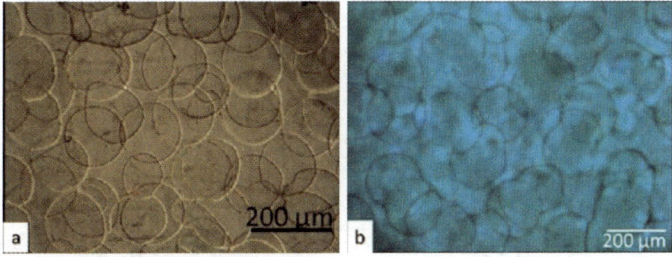

Figure 1.9 (a) An array of interconnected silver nanoparticle rings; (b) a four-layer electroluminescent device built by depositing on PET a silver ring array and ZnS and BaTiO₃ layers by screen printing followed by the deposition of a continuous layer of silver nanoparticles as electrode. (Reprinted with permission from M. Layani *et al.*, *ACS Nano*, 2009, **3**, 3537. Copyright 2009 American Chemical Society.)

shown in **Figure 1.8**. It is important to notice that the outer part of the ring appears denser with respect to the inner side and that the particles are well packed but not ordered. We will see later in this chapter that behind this specific behavior is another hidden secret of the coffee-stain effect. The top and side SEM images (**Figure 1.8(c)** and **(d)**) show clearly the high packing degree to which the nanoparticles are pushed by the capillary flow to the droplet edge.

The high conductivity exhibited by silver nanoparticle-packed rings opens the door for fabricating an interconnected two-dimensional (2D) array with properties that can be conductive and transparent[19], which is very interesting for a wide range of applications from electroluminescent devices to solar cells. An array of silver rings can be obtained by ink-jet printing using a water solution of silver nanoparticles as the ink. Every droplet leaves a silver ring due to the coffee-stain effect and a layer with properties that are close to those of indium−tin oxide (ITO) is produced by repetitive writing (**Figure 1.9**). The fabrication of such transparent and conductive coatings requires a sufficient fraction of the rings to be electrically interconnected and that the width of the rim rings does not exceed 10 μm, which is lower than the threshold of resolution to the naked eye[20]. The advantage of this process, besides fabricating optically transparent layers using a simple method is the possibility of deposition on flexible substrates. The low resistance of the 2D array is not affected by bending the substrate at angles below 20°.

 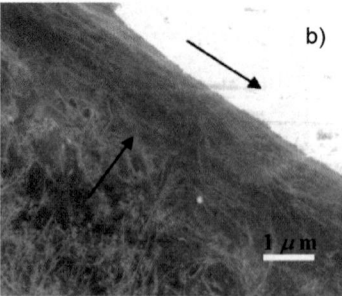

Figure 1.10   (a) Scheme of self-organization of CNTs during the droplet-drying
process; (b) SEM image which shows the alignment of SWNTs at the
outmost edge of the drop and transition between the two domains.
(Reprinted with permission from Q. Li *et al.*, *J. Phys. Chem. B*, 2006,
110, 13926. Copyright 2006 American Chemical Society.)

1.4 SHAPE MATTERS

We have seen several examples of how the coffee-stain effect can be
used for obtaining ordered rings of self-assembled packed particles of
different types. Most of the cases are restricted to spherical particles of
different dimensions and compositions, but what about shape effects?
The first case for discussion is the evaporation of a droplet deposited
from a stable dispersion of carbon nanotubes (CNTs). CNTs in solution
have shown a tendency to align on surfaces under a flow[21] and in the
case of an evaporating droplet, which is pinned at the contact angle,
the capillary flow pushes the CNTs toward the droplet edge but the
anisotropic interaction and the reduced angular mobility tend to align
CNTs in the flow direction. However, when they approach the contact
line, CNTs are forced to align parallel to the edge by changing their
flow directions due to geometrical constraints (**Figure 1.10**). At room
temperature, therefore, on a wettable surface such as glass or silica
wafer, a redistribution, accumulation, and organization of CNTs along
the edge of the droplet can be observed[22]. With respect to multi-wall
nanotubes, single-wall nanotubes show a higher tendency to self-
organize at the droplet edge; in fact, single-wall nanotubes are more
flexible which results in easier bending and they are able to adjust their
orientations at the outer edge; to gain higher ordering, and at a larger
extent.

  If we move from the case of nanotubes to colloidal particles then
things become even more intriguing. The shape plays an important role
at the interface[23] simply because the long-range attraction between the
anisotropic particles is more than two orders of magnitude stronger
than between spherical objects[24]. This means that at the liquid−air

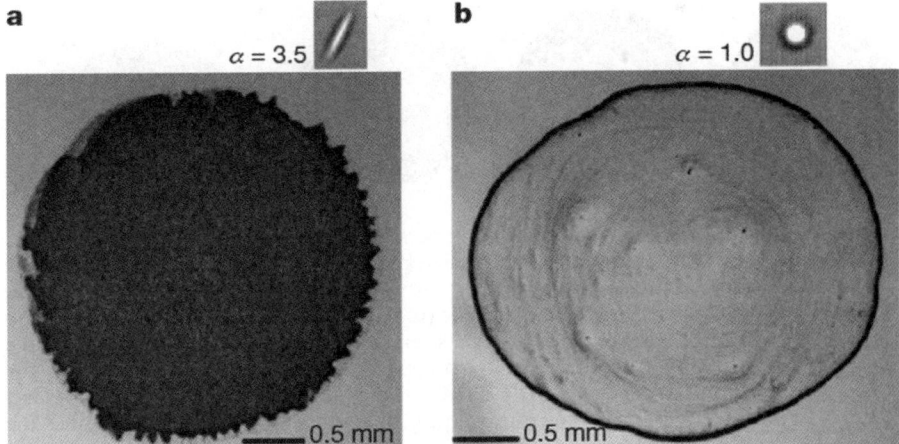

Figure 1.11 Optical images after evaporation on a glass substrate of a colloidal droplet of polystyrene particles of different shapes (a) ellipsoids with major axis to minor axis ratio, $\alpha = 3.5$ and (b) spherical particles, $\alpha = 1$. (Reprinted by permission from Macmillan Publishers Ltd: *Nature*, 2011, **476**, 308. Copyright 2011.)

interface of a drying droplet the higher interaction of elongated objects is expected to affect the coffee-stain effect, which is actually the case. The strength of the attractive interparticle interaction between ellipsoids can be strong enough to counterbalance the forces that drive spherical particles towards the edge of the evaporating droplet suppressing, therefore, the coffee stain[25]. Yunker and co-workers, have repeated the experiment of Deegan with polystyrene microspheres using elongated particles and the result is quite striking as a uniform coating, instead of a ring, is obtained by changing the particle shape. **Figure 1.11** shows the images of the final deposition in the case of ellipsoidal particles (**Figure 1.11(a)**) or microspheres (**Figure 1.11(b)**), with a uniform distribution for ellipsoids and a ring for the spheres. While spherical particles are transported to the droplet edge *via* capillary flow, leaving a ring after evaporation, in the case of anisotropic particles (with α, the major axis/minor axis ratio, >1.2), these move to the edge only until they reach the air−water interface. At this point the long range attraction between the elongated particles favors the formation of loosely packed aggregates at the air−water interface. Breaking or moving these clusters will cost energy and their mobility is reduced in comparison with spherical particles and the strength of the radial outward flow, which causes the accumulation of particles at the droplet edge, is not enough to move these clusters from the interface (see

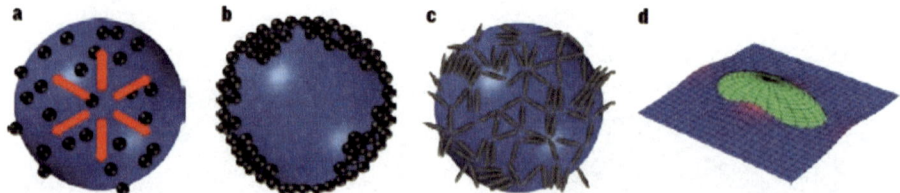

Figure 1.12   Preventing the coffee-ring effect. (a) During drying of a droplet (blue; top view) containing suspended particles (black spheres) the liquid flow due to the coffee stain effect from the interior towards the edge drives the particle to the contact line (red arrows). (b) The particles accumulate near the edge and form a stain. (c) This coffee stain effect can be prevented by getting self-assembly of ellipsoidal particles (grey) at the liquid–air interface. (d) The local curvature of the ellipsoidal particle's surface (green) and the wetting of the particle by the fluid force the interface (blue and pink) to become deformed, pulling the liquid up near the middle of the ellipsoid and pushing it down near its tips. (Reprinted by permission from Macmillan Publishers Ltd: *Nature*, 2011, **476**, 286. Copyright 2011.)

Figure 1.12)[26]. If evaporation is repeated by adding a small amount of surfactant (sodium dodecyl sulfate) to a suspension of elongated particles, the coffee-stain effect is restored. The surfactant reduces the surface tension of the droplet and deformations at the interface will be of a shorter range and will require less energy (**Figure 1.13**).

This shape effect can be used in a more general way to control the coffee-stain effect in evaporating droplets. The addition of small

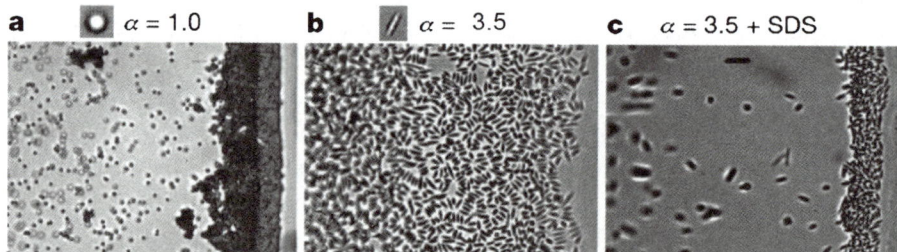

Figure 1.13   Microscope images of a region within 40 mm from the droplet contact line. The images have been taken in the middle of the evaporation process (time $t/t_{final} = 0.5$), for three different types of colloidal suspensions. (a) Spherical particles, $\alpha = 1$; (b) ellipsoids ($\alpha = 3.5$), and (c) ellipsoids ($\alpha = 3.5$) mixed with surfactant (sodium dodecyl sulfonate (SDS); 0.2 wt%). The spheres form a ring with dense packing at the contact line whilst ellipsoids form loosely packed structures; addition of surfactant in the suspension, which contains the ellipsoids lowers the drop surface tension and pushes the particles again to pack closely at the contact line. (Reprinted by permission from Macmillan Publishers Ltd: *Nature*, 2011, **476**, 308. Copyright 2011.)

numbers of ellipsoids to sphere suspensions can also suppress the ring formation. The capability of suppressing the ring formation depends of the relative sizes of the two types of particles when only the diameter of the spheres is larger than the minor axis of the ellipsoids does the coffee ring not form.

1.5 THE DIMENSIONS ALSO MATTER

We have just seen that the shape of the particles assembling at the contact line plays a very important role, but what about the dimensions of these particles? Let's start with the simple consideration that at the contact line the droplet edge has a very well defined profile. The thickness of the droplet meniscus gradually increases moving toward the center of the contact line. When approaching the contact line, the particles stop before the edge at a distance that is dependent on the matching point between the particle size and the local meniscus. If the solution contains particles of different dimensions, the smaller particles will move closer to the external rim of the droplet[27], which in turn separates into another distinct ring. The meniscus induced phase separation process is schematically illustrated in **Figure 1.14**. The result is a size-dependent physical particle partition with interesting possibi-

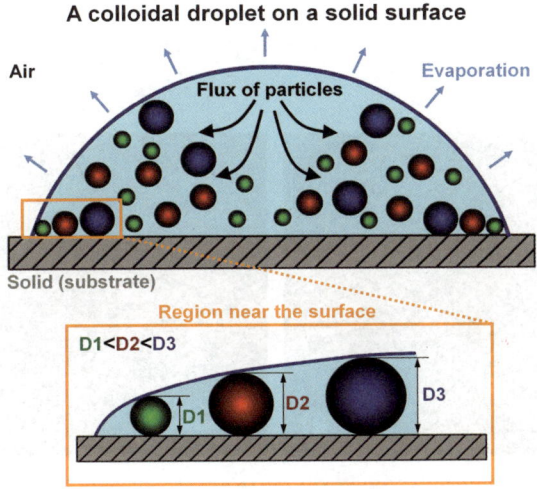

Figure 1.14   Schematic illustration of the concept of size-dependent particle separation, which is observed close to the contact line in an evaporating droplet containing particles of three different sizes. The smallest particles separate closer to the contact line. (Reprinted with permission from T.-S. Wong *et al.*, *Anal. Chem.*, 2011, **83**, 1871. Copyright 2011, American Chemical Society.)

lities for applications as a nanochromatographic device[28] for separation
and concentration of molecular/cellular components in biological fluids.
Figure 1.15(a) shows the optical fluorescence image, with the particle
separation after evaporation into three concentric but well defined
domains of negatively charge-stabilized polystyrene fluorescent particles
of 40 nm (green), 1 µm (red), and 2 µm (blue). The feasibility of the
method for nanochromatography has been demonstrated as a proof-of-
concept using a mixture of fluorescently-labelled antimouse IgG
antibodies (<10 nm), *Escherichia coli* expressing a green fluorescent
protein (order of 500 nm), and fluorescently-labelled murine B-lymphoma cells (WEHI-231) (order of 5 µm) suspended in deionized water at
predetermined concentrations. The different components separate
effectively as a function of their physical size from the smaller antimouse
IgG antibodies (blue) at the outermost part of the ring **(Figure 1.15b)**,
followed by *E. coli* (green) that formed the second rim and the B-
lymphoma (red) cells that deposited in the third inner part of the ring. All
these rings appear to be very well separated. During the evaporation of a
droplet containing particles, molecules or micro-organisms the coffee-
stain effect can, therefore, be used as a tool for separation and
concentration.

The same size effect that we have just described also has a direct
influence on the final diameter of the dried droplet. This is quite
intuitive because the shape of the contact line does not allow the larger
particles to move so close to the droplet edge[29]. The diameter of the

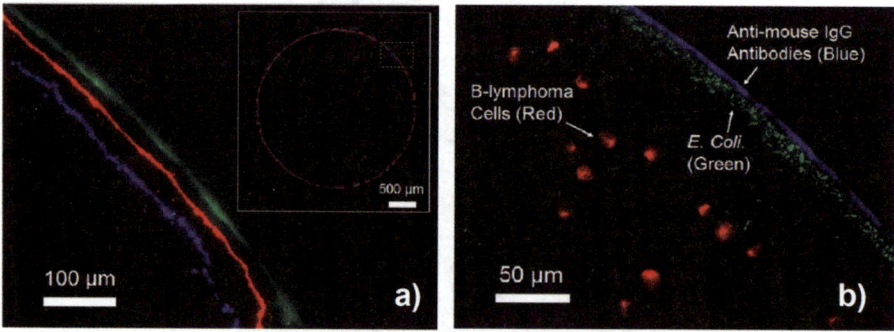

Figure 1.15   (a) Optical fluorescence image showing the separation of 40 nm (green),
1 µm (red), and 2 µm (blue) negatively charge-stabilized polystyrene
fluorescent particles after evaporation. (b) Optical fluorescence image
showing the separation of antimouse IgG antibodies (blue), *Escherichia
coli* (green), and B-lymphoma cells (red) in a dried liquid drop.
(Reprinted with permission from T.-S. Wong *et al.*, *Anal. Chem.*, 2011,
83, 1871. Copyright 2011, American Chemical Society.)

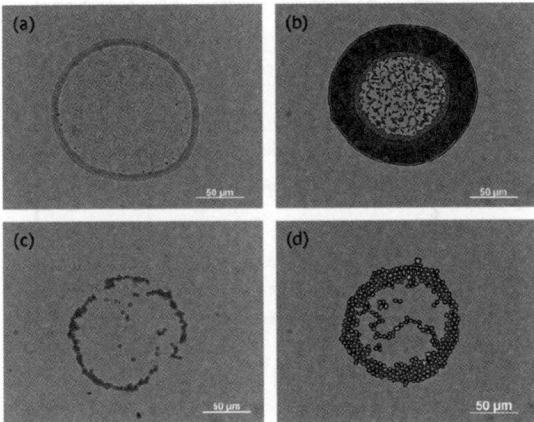

Figure 1.16 Bright field microscopy images of ink-jet printed droplets obtained by aqueous suspensions containing (a) 0.33 μm, (b) 1 μm, (c) 3 μm and (d) 5 μm sized silica particles. The photograph (c) has been focused on the contact line rather than on the particles. (Reprinted with permission from ref. 29.)

dried droplet increases, therefore, with a decrease of the particle size because smaller particles can move closer to the contact line. A demonstration of this has been obtained by depositing, *via* ink-jet printing, droplets of equal volume, which can be exactly controlled with this technique, containing silica microspheres of different dimensions. The result of the experiment is shown in **Figure 1.16**. The images of dried droplets with silica particles of (a) 0.33 μm, (b) 1 μm, (c) 3 μm and (d) 5 μm indicate that smaller particles leave larger rings. The shape of the droplet modulates the movement of the particles in the direction of the contact line with preferential deposition of smaller particles closer to the contact line.

When we talk about dimensions we should not forget that the droplet itself has to be considered and the question to ask is if the coffee-stain observation is dependent on the droplet size or not. The answer is that actually it is. There exists a minimum droplet size to form a coffee-stain structure. If the droplet is smaller than the critical diameter, d_c, the particles tend to form a homogeneous layer instead of a ring[30]. The value of d_c depends on several parameters: (i) relative humidity (RH), at higher RH values d_c decreases; (ii) particle concentration, d_c decreases with the increase of concentration and (iii) particle dimension, the pinning effect is reduced with decreasing particle size[31]. We have noticed that the dimension of the droplet changes the evaporation rate. In general, smaller droplets should evaporate faster. How does this affect the critical size? It is possible to answer this question by

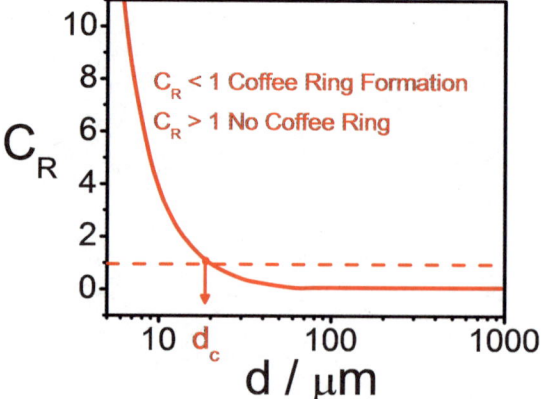

Figure 1.17 A semilog plot of C_R obtained by a model of an evaporating droplet of 100 nm particles at a fixed concentration. The critical droplet diameter, d_c, corresponds to $C_R = 1$, *i.e.* $\tau_{evap} = \tau_{particle}$. (Reprinted with permission from X. Shen *et al.*, *J. Phys. Chem. B*, 2010, **114**, 5269.)

comparing the relative time scales of the two phenomena: the solvent evaporation time, τ_{evap}, and the time necessary for two particles to keep in contact close to the droplet edge, $\tau_{particle}$. The ratio between $\tau_{particle}$ and τ_{evap} is defined by the dimensionless number, C_R:

$$C_R = \frac{\tau_{particle}}{\tau_{evap}} \tag{1.2}$$

Figure 1.17 shows a semilog plot of C_R obtained by a model of an evaporating droplet of 100 nm particles at a fixed concentration. The critical droplet diameter corresponds to $C_R = 1$, *i.e.* $\tau_{evap} = \tau_{particle}$. For $C_R > 1$, no coffee ring structures are formed because the evaporation time is too fast to allow for the formation of a pinning layer at the contact line. When $C_R < 1$, the nanoparticles have enough time to migrate to the contact line and pinning and formation of the ring is finally observed. The droplet size, therefore, should be larger than a critical value that allows the liquid evaporation time to be smaller than the migration time of the particles. Therefore, the condition $C_R \ll 1$ is necessary for observing ring formation.

This is not all about the importance of the particle's dimension in drying droplets, and some other surprises can be found behind the evaporation of a colloidal solution. We have learnt that the evaporating flow, if Marangoni flow is suppressed, drives the particles to accumulate toward the contact line, with a motion from the center to the droplet edge. It is interesting to observe the particles that reach

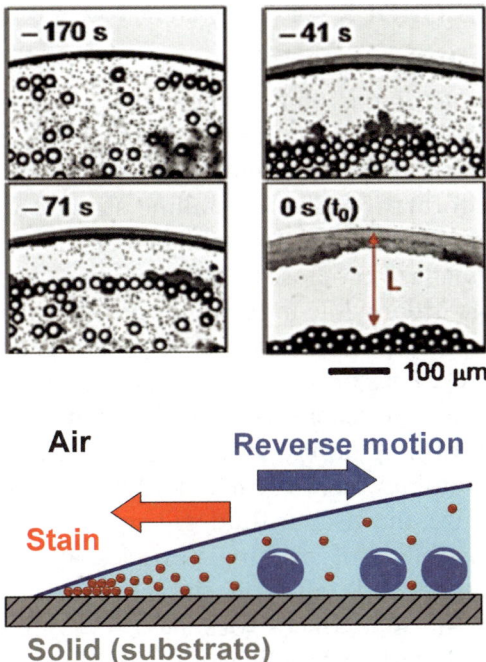

Figure 1.18 Top figure: top-view of real time optical micrographs which show the normal and reverse motions of 2 and 20 μm diameter particles, respectively, in an evaporating water droplet of 1.5 mm diameter. Bottom figure: schematic view of the motion of the particles of different size in an evaporating droplet. (Adapted with permission from B. M. Weon *et al.*, *Phys. Rev. E*, 2010, **82**, 015305(R).)

the external part of the droplet and then reverse to return to the droplet center[32]. This phenomena has been observed when the dimension of the particles are such a size that the lateral immersion capillary forces that depend on the particle dimension and contact angle[33] are factored in. Polystyrene particles of 20 μm in diameter have been found to reverse the motion when they approach the contact line. The reverse motion is preceded by the formation of the coffee-stain effect at the droplet edge.

The effect of particle size becomes clear when the solution contains polystyrene microspheres of different dimensions (2 and 20 mm) that are left to evaporate simultaneously. The smaller particles accumulate at the droplet edge while the larger ones reverse the motion (**Figure 1.18**). If the outward force generated by the capillary force near the contact line is greater than the inward flow of the coffee stain effect the particles invert the motion.

1.6   ACHIEVING ORDER AT THE CONTACT LINE

We have seen that the capillary flow during the evaporation of a droplet, if the contact line is pinned and the Marangoni effect is weak, produces an accumulation of matter at the droplet edge. The density of the particle packing can be very high, but what about true ordering in the rings, *i.e.* the formation of a crystalline structure?[34,35] There is an interesting experimental finding to emphasis before getting to the root of the problem, which is very close to the droplet edge, the particle array shows a long-range order while in the inner part of the ring the particles tend to be less densely packed. This observation suggests that during evaporation some changes in the flow rate may occur. Transitions from disorder to order are, in general, kinetically controlled processes. The cooling of a glass-forming liquid gives a crystalline solid or a glass as a function of the cooling rate. If this is slow enough for atoms or molecules to organize, the viscous fluid will crystallize and form a glassy structure. The possibility of packing particles at the contact line in ordered structures depends also on the kinetics of this process, as demonstrated by Marin and co-workers[36]. They have shown in an experiment realized with monodispersed red-fluorescent polystyrene particles (0.5–2 μm diameter) that the deposition rate has two main regimes, a slow one at the beginning of the process and a second one that is faster in the last phase of evaporation, in accordance with similar observations[37]. A slow liquid flow velocity allows ordered packing of the particles while during the faster one the particles form a disordered array. The external side of the ring (SEM images in **Figure 1.19**) has an ordered crystalline structure but a clear transition to a disordered arrangement of the particles is observed towards the center of the droplet. The top view of the image also shows that the ordered region is formed by different particle arrangements, a hexagonal (the outermost lines), a square (middle) and finally again a hexagonal array. The origin of this order−disorder transition can be understood if the radial particle velocity *versus* time is analyzed; the theoretical model of Deegan[7,9] says that close to the contact line, the height-averaged radial velocity, u, depends on time, t, and distance, r, from the center of the drop[38]:

$$\vec{u}(r,t) = \frac{2\sqrt{2}D_{va}\Delta c}{\pi\rho} \frac{1}{\theta(t)} \frac{1}{\sqrt{R(R-r)}} \qquad (1.3)$$

where R is the radius of the droplet, D_{va} the diffusion constant of vapor in air, Δc, the vapor concentration difference between the drop surface and the surrounding, ρ, the liquid density, $(R-r)^{-1/2}$, the spatial

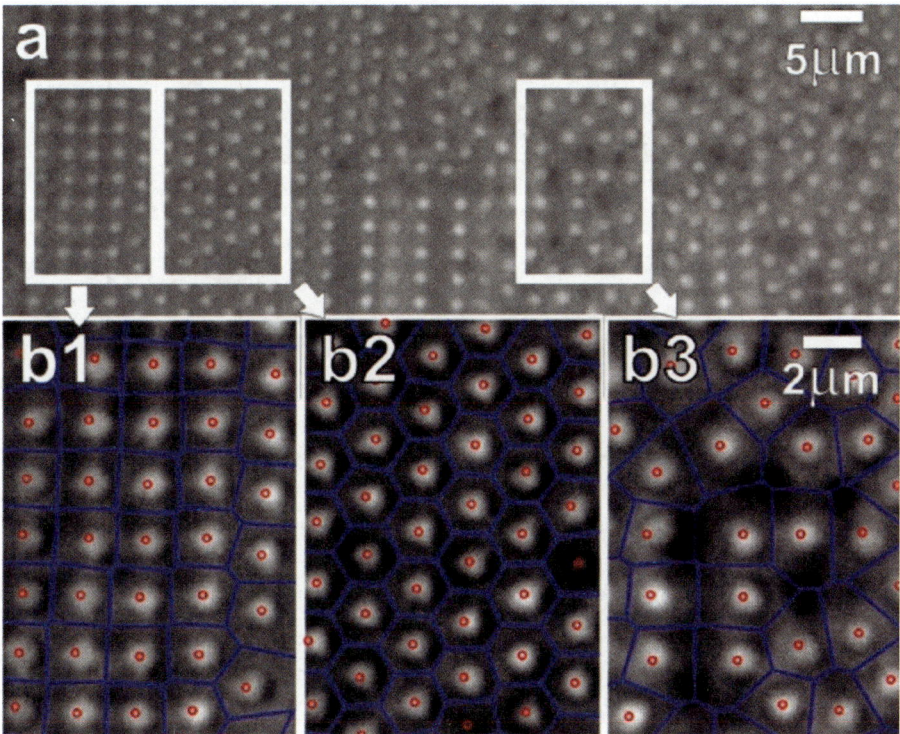

Figure 1.19 (a) SEM top image of different patterns observed in the ring, the first layers from the contact line (left side of the figure) are not shown. The details of the patterns sequence in the ring are shown in (b1)−(b3): (b1) square packing close to the contact line, (b2) hexagonal packing, and (b3) disordered packing. (Reprinted with permission from Á. G. Marín *et al.*, *Phys. Rev. Lett.*, 2011, **107**, 085502.)

divergence and $\theta(t)$, the contact angle. This means that the fluid speed does not change only as a function of position from the center of the droplet but also with the evaporation time. The origin of the temporal divergence is the vanishing contact angle and this term is responsible for the sudden change in fluid velocity that produces the order-disorder transition. The experimental results of Marin *et al.*[36] show that (**Figure 1.20**) when the evaporating process is entering the final stages there is dramatic increase in the velocity that corresponds to a disordered packing of particles. This sudden change is a kind of 'rush hour' moment. The particles that arrive at the beginning of the evaporation at a low deposition rate have enough time to rearrange by Brownian motion and reach an ordered packing while those arriving later, during the rush hour when the flow speed is much higher, do not have enough time to organize. It is a kind of Tetris® effect; at the

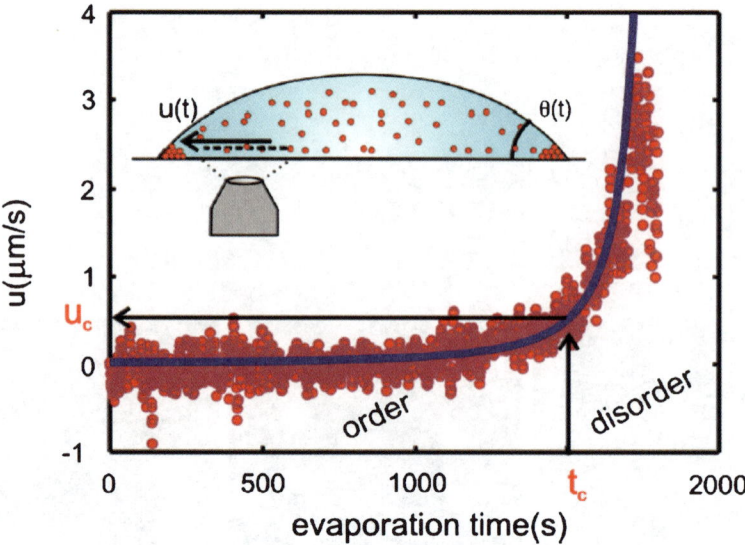

Figure 1.20   Plot of the radial particle velocity *versus* time; the solid blue line shows
fluid velocity predicted by the model. The inset shows the direction of
the radial velocity in the droplet. The order to disorder transition time
is the critical time, t_c, with the corresponding critical velocity u_c.
(Reprinted with permission from Á. G. Marín, *et al.*, *Phys. Rev. Lett.*,
2011, **107**, 085502.)

beginning the pieces arrive at a slow rate but as the game progresses, the
rate at which the pieces arrive increases and it becomes more difficult to
place the blocks in ordered layers.

Beside the circular shape, the stain evaporation produces a sequential
arrangement of different packing lattices; starting from the edge and
moving towards the stain centre, in fact, a transition from square to
hexagonal particle organisation can be observed. This specific sequence
is not unexpected[39]. While in an open system the most efficient lattice
arrangement is the dense hexagonal structure, an evaporating droplet is
a confined system because the particles remain restricted between the
substrate and the liquid−vapor interface. The most efficient packing in
such a confined system gives a sequence of square and hexagonal
structures[40,41]. At the contact angle rim the new layer has square
packing, while with the increase of the space available moving away
from the edge a denser hexagonal structure can form.

Achieving order at the contact line is quite a general phenomenon
and different types of particles show crystalline packing such as
mesoporous silica beads (**Figure 1.21**). Rings of mesoporous particles

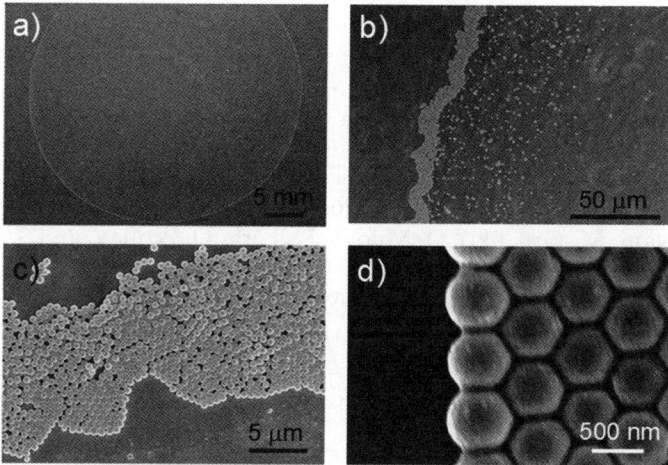

Figure 1.21 2D close packed layer of mesoporous silica microbeads at the contact line obtained by self-assembly under slow evaporation rate of the solvent. (Reprinted from L. Malfatti *et al.*, *Microporous Mesoporous Mater.*, **163**, 356. Copyright 2012, with permission from Elsevier.)

have the advantage of forming a porous rim with the capability of targeting specific molecules[42].

REFERENCES

1. P. Innocenzi, L. Malfatti, M. Piccinini, D. Grosso, and A. Marcelli, *Anal. Chem.*, 2009, **81**, 551.
2. S. Maenosomo, C. D. Dushkin, S. Saita, and Y. Yamaguchi, *Langmuir*, 1999, **15**, 957.
3. T. Kajiya, D. Kaneko, and M. Doi, *Langmuir*, 2008, **24**, 12369.
4. D. Kaya, V. A. Belyi, and M. Muthukumar, *J. Chem. Phys.*, 2010, **133**, 114905.
5. D. Soltman and V. Subramanian, *Langmuir*, 2008, **24**, 2224.
6. R. Dou, T. Wang, Y. Guo, and B. Derby, *J. Am. Ceram. Soc.*, 2011, **94**, 3787.
7. R. D. Deegan, O. Bakajin, T. F. Dupont, G. Huber, S. R. Nagel, and T. A. Witten, *Nature*, 1997, **389**, 827.
8. R. D. Deegan, *Phys. Rev. E*, 2000, **61**, 475.
9. R. D. Deegan, O. Bakajin, T. F. Dupont, G. Huber, S. R. Nagel, and T. A. Witten, *Phys. Rev. E*, 2000, **62**, 756.
10. K. Kochiya and I. Ueno, *Ann. N. Y. Acad. Sci.*, 2009, **1161**, 234.
11. R. Bhardwaj, X. Fang, P. Somasundaran, and D. Attinger, *Langmuir*, 2010, **26**, 7833.
12. H. Hu and R. G. Larson, *Langmuir*, 2005, **21**, 3963.
13. B. J. Fischer, *Langmuir*, 2002, **18**, 60.

14. H. Hu and R. G. Larson, *J. Phys. Chem. B*, 2002, **106**, 1334.
15. A. S. Sangani, C. H. Lu, K. H. Su, and J. A. Schwarz, *Phys. Rev. E*, 2009, **80**, 011603.
16. H. Hu and R. G. Larson, *J. Phys. Chem. B*, 2006, **110**, 7090.
17. H. Hu and R. G. Larson, *Langmuir*, 2005, **21**, 3972.
18. S. Magdassi, M. Grouchko, D. Toker, A. Kamyshny, I. Balberg, and O. Millo, *Langmuir*, 2005, **21**, 10264.
19. M. Layani, M. Gruchko, O. Milo, I. Balberg, D. Azulay, and S. Magdassi, *ACS Nano*, 2009, **3**, 3537.
20. J. Perelaer, P. J. Smith, M. M. P. Wijnen, E. van den Bosch, R. Eckardt, P. Ketelaars, and U. S. Schubert, *Macromol. Chem. Phys.*, 2009, **210**, 387.
21. H. Xin and A. T. Woolley, *Nano Lett.*, 2004, **4**, 1481.
22. Q. Li, Y. T. Zhu, I. A. Kinloch, and A. H. Windle, *J. Phys. Chem. B*, 2006, **110**, 13926.
23. B. Madivala, J. Fransaer, and J. Vermant, *Langmuir*, 2009, **25**, 2718.
24. B. J. Park and E. M. Furst, *Soft Matter*, 2011, **7**, 7676.
25. P. J. Yunker, T. Still, M. A. Lohr, and A. G. Yodh, *Nature*, 2011, **476**, 308.
26. J. Vermant, *Nature*, 2011, **476**, 286.
27. V. H. Chhasatia and Y. Sun, *Soft Matter*, 2011, **7**, 10135.
28. T.-S. Wong, T.-H. Chen, X. Shen, and C.-M. Ho, *Anal. Chem.*, 2011, **83**, 1871.
29. J. Perelaer, P. J. Smith, C. E. Hendriks, A. M. J. van den Berg, and U. S. Schubert, *Soft Matter*, 2008, **4**, 1072.
30. X. Shen, C.-M. Ho, and T.-S. Wong, *J. Phys. Chem. B*, 2010, **114**, 5269.
31. T. Ondarcuhu and A. Piednoir, *Nano Lett.*, 2005, **5**, 1744.
32. B. M. Weon and J. H. Je, *Phys. Rev. E*, 2010, **82**, 015305, (R).
33. P. A. Kralchevsky and N. D. Denkov, *Curr. Opin. Colloid Interface Sci.*, 2001, **6**, 383.
34. A. P. Sommer, M. Ben-Moshe, and S. Magdassi, *J. Phys. Chem. B*, 2004, **108**, 8.
35. A. P. Sommer and N. Rozlosnik, *Cryst. Growth Des.*, 2005, **5**, 551.
36. Á. G. Marín, H. Gelderblom, D. Lohse, and J. H. Snoeijer, *Phys. Rev. Lett.*, 2011, **107**, 085502.
37. C. Monteux and F. Lequeux, *Langmuir*, 2011, **27**, 2917.
38. Y. O. Popov, *Phys. Rev. E*, 2005, **71**, 036313.
39. M. Abkarian, J. Nunes, and H. Stone, *J. Am. Chem. Soc.*, 2004, **126**, 5978.
40. P. Pieranski, L. Strzelecki, and B. Pansu, *Phys. Rev. Lett.*, 1983, **50**, 900.
41. B. Pansu, Pi. Pieranski, and Pa. Pieranski, *J. Phys.*, 1984, **45**, 331.
42. L. Malfatti, Y. Tokudome, K. Okada, S. Yagi, M. Takahashi, and P. Innocenzi, *Microporous Mesoporous Mater.*, 2012, **163**, 356.

CHAPTER 2

The Tears of Wine. The Marangoni Effect, a Fluid Phenomenon for Self-Assembly and Organization

Another example of a chemical−physical phenomenon that can be successfully applied for making self-organized structures in materials is the Marangoni effect. This is a fluid phenomenon that gives rise to popular effects such as the so-called 'tears of wine'. This effect is also observed during the deposition of a film from a liquid phase or an evaporating droplet[1] and is used for producing sub-micrometer ordered structures.

2.1 WINE TEARS AND THE MARANGONI EFFECT

In Chapter 1 we discussed how an evaporating liquid droplet can become an unexpected tool for self-assembly. The coffee-stain 'mystery' has been used to open the door to new routes for producing self-ordered structures. However, before a coffee it is better to have, especially with the approach of dinner, an '*aperitivo*' which means a good glass of wine. However, what is the connection between wine and self-assembly? Again the answer places attention on the evaporation phenomena; wine contains alcohol, which during evaporation can easily trigger some 'strange' chemical−physical phenomena. A simple experiment can be done at home to observe this 'miracle', which is connected with wine. Pour some wine into a glass and carefully observe what happens as the alcohol evaporates. The wine should start to go up along the surface of the glass forming a thin film. As time elapses the wine accumulates forming a thicker rim at the top of the film and some droplets, under their own weight, slip back into the wine. These droplets, which

Water Droplets to Nanotechnology: A Journey Through Self-Assembly
By Plinio Innocenzi, Luca Malfatti and Paolo Falcaro
Published by the Royal Society of Chemistry, www.rsc.org

continuously form and fall into the wine are known as the 'tears of wine' and behind this macroscopic phenomenon there is the so-called Marangoni effect. **Figure 2.1** shows the wine tears that form on the surface of a wine glass.

The phenomenon that is the 'tears of wine' was described for the first time[2] by the Italian physicist Carlo Marangoni[3] and is know as the Marangoni effect. We have already seen in Chapter **1** that the Marangoni effect is responsible for reversing the capillary flow of a coffee stain, but this fluid phenomenon is more general and complex. The Marangoni flow affects the coating technologies for the deposition

Figure 2.1   The 'tears of wine' in a glass induced by alcohol evaporation (picture taken by L. Malfatti using a *Canonau* wine. The glass has been kindly provided by *Papallero* wine-bar.).

of films from liquid phases, welding, electron beam melting of metals, crystal growth, and stabilizes soap bubbles by giving them elasticity. We can describe the Marangoni effect in a general sense as follows: the formation of a concentration or temperature gradient along the surface of a liquid which induces a surface tension gradient that drives the flowing process. How can this process explain the tears of wine? The wine in the glass can be any kind of liquid mixture, which upon evaporation, shows a tendency to form concentration gradients on the surface, wets the wall forming a meniscus[4] and a thin liquid film spontaneously rises above the wine reservoir on the glass wall pulled by capillary forces.[5] Evaporation in this thin film is relatively fast and mainly due to alcohol as it is more volatile than water. The concentration of the alcohol will also decrease as a function of the height, z, from the wine reservoir.[6] If we consider, as an example, the simpler case of an ethanol−water mixture, because of the lower surface tension, γ, of ethanol with respect to water, the liquid surface tension increases with the decrease of ethanol in the mixture as shown in **Figure 2.2**. This means that a gradient of alcohol concentration, $\partial C/\partial z$, is created along the film surface as a function of the distance z from the wine reservoir. The concentration gradient in turn produces a surface tension gradient, γ, which is the force that drives the spreading of the liquid on the glass wall. This can be written as:

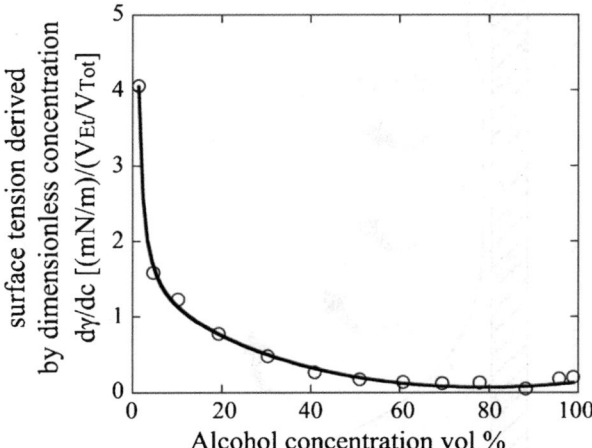

Figure 2.2 The change of surface tension, $d\gamma/dc$, of a water−ethanol mixture as a function of ethanol fraction. The fraction, c, is V_{Et}/V_{Tot}, which is the ethanol volumetric fraction. (Reprinted with permission from R. Tadmor *J. Colloid. Interface Sci.*, **332**, 854. Copyright 2009 with permission from Elsevier.)

$$\frac{\partial \gamma}{\partial z} = \left(\frac{\partial \gamma}{\partial C} \right) \left(\frac{\partial C}{\partial z} \right) \qquad (2.1)$$

The result is a surface tension gradient along the glass wall where surface tension increases with increasing distance from the liquid reservoir. The process is shown in **Figure 2.3**. Finally, therefore, the consequence of the formation of a surface tension gradient upon evaporation (Marangoni effect) is the rise of the liquid on the glass wall. It is important to keep in mind that the higher the alcohol content in the wine so the 'wine tear' will be more pronounced.

The tendency of a fluid to exhibit convection flows due to the Marangoni effect depends on several parameters, such as viscosity, η, and is expressed by the Marangoni number, Ma, a dimensionless number which is defined by:

$$Ma = \frac{\left(\frac{\partial \gamma}{\partial T} \right) t \nabla T}{\eta \alpha} \qquad (2.2)$$

where the temperature gradient dominates the effect. γ is the surface tension, $\partial \gamma / \partial T$ the temperature derivative of the surface tension, t the

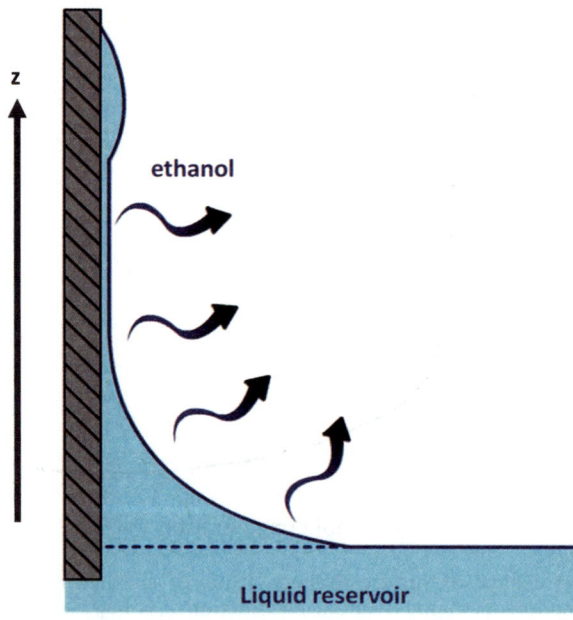

Figure 2.3   The formation of wine tears in a glass cup.

liquid thickness, α the thermal diffusivity of the solution, and ∇T the gradient of temperature near the liquid surface:

$$Ma = \frac{\left(\frac{\partial \gamma}{\partial C}\right) t \nabla C}{\eta D} \qquad (2.3)$$

where the concentration gradient is the driving force, with ∇C the gradient of concentration near the liquid surface, D the mass diffusivity, $\partial \gamma / \partial C$ the concentration derivative of the surface tension. When the viscosity of the liquid and its diffusivity are low enough, the gradient in surface tension gives rise to an instability of the fluid so that the flow symmetry and steadiness are lost. This threshold point defines the so-called critical Marangoni number, Mac.

2.2 SELF-ORGANIZATION DURING FORMATION OF A LIQUID FILM *VIA* DIP-COATING AND SPIN-COATING

The Marangoni effect in thin-film technologies is a potential source of problems because convective flow can produce defects such as uncontrolled striations. Striations are defects that appear as radial ridges and thickness undulations that point directly along the flow direction during spin-coating[7] or the withdrawal direction in the case of dip-coating.[8] In the case of spin-coated films, such as photoresists, sol−gel and polymer coatings, radiative striations, which could appear upon deposition, have to be carefully avoided with proper processing technology.[9,10] The formation of striations during film deposition can, however, be turned into a feasible route to self-organization if the phenomenon is controlled somehow. Film processing can be designed to produce controlled patterns, in terms of dimension and alignment, without further processing. An example that has been well studied is the spin-coating of sol−gel films. In general, the precursor sol contains a solvent (alcohol), water for hydrolysis, and as a by-product, a catalyst and an alkoxide as the oxide network former. The evaporation of the solvent triggers the transition from sol to gel, which during film deposition is very fast. Instability in the solvent evaporation produces striation in the film.[11,12] The origin of this instability is a random fluctuation in surface composition, which in turn causes local surface tension gradients. The process is depicted in **Figure 2.4**, which shows the capillary instability that arises during the drying stage in spin coating.[8] The evaporation process generates a solvent-depleted surface layer that has a different surface tension with respect to the solution beneath it.

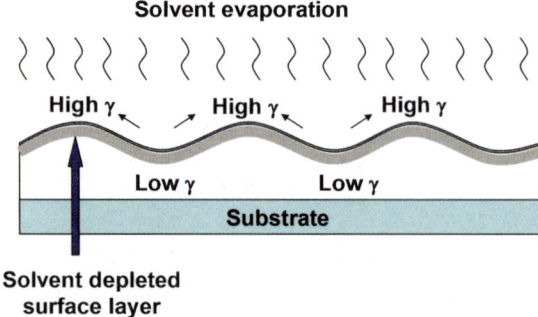

Figure 2.4   Regular striations during the drying stage of spin coating forming due to the Marangoni effect. A solvent-depleted surface layer with a higher surface tension with respect to the solution beneath can arise during evaporation. The local instabilities are amplified and generate gradients in surface tensions which produce capillary-driven motions of fluid at the surface. (Adapted with permission from D. J. Taylor *et al., Chem. Mater.,* 2002, **14**, 1488. Copyright 2002, American Chemical Society.)

Local random variations in compositions, however, can cause differences in surface tensions (the zones with high and low γ) to arise that trigger a lateral fluid motion, which is responsible for the striation patterns. Because the main driver of this instability is a compositional change, more than a temperature gradient, the choice of solvent will either suppress or strongly reduce the striation.

The formation of these structures, however, can be also seen as an opportunity to form self-organized patterns in films, if the Marangoni effect is properly managed. An example is the formation of linearly-aligned striations in crystalline titania films obtained *via* sol−gel dip-coating. Linear patterns spontaneously align in the direction of the substrate withdrawal as shown in **Figure 2.5**. The height and the spacing of such patterns depend on the thickness of the film, the viscosity of the solution and the distance from the film edge. A similar parallel alignment can be reproduced *via* spin casting of polymer blends. In this case, the Marangoni effect combines with a phase separation mechanism to give aligned structures enriched in polystyrene or in poly (vinylpyrrolidone).[13]

The formation of aligned patterns in polymer films *via* the Marangoni effect can be exploited for producing ordered alignments of nanostructures.[14] The instability in a spin-cast film obtained by a silver-nanowire polymer (poly(2-vinylpyridine)) suspension triggers the formation of alternating polymer/nanowire striation patterns. The silver

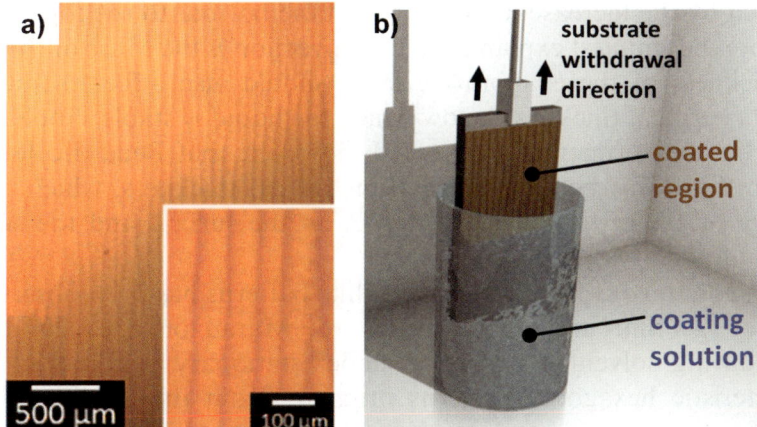

Figure 2.5   (a) Optical image of self-organized patterns of titania aligned in a direction parallel to substrate withdrawal; (b) formation of the striations during dip-coating. (Adapted with permission from H. Uchiyama *et al.*, *J. Phys. Chem. C*, 2012, **116**, 939. Copyright 2012, American Chemical Society.)

nanowires are transported toward the concave regions to balance the non-uniform distribution of the surface tension in the polymer solution.

2.3   BÉNARD CELLS AND SELF-ASSEMBLY

Linear striation is not the only defect that can be observed as a consequence of the Marangoni effect during the processing of thin films. There is in fact another peculiar structure that can form. We have just described the case of evaporating systems that are under dynamic conditions, such as rotation during spin-coating and pulling in dip-coating but what happens in a fluid that is stationary? In thin layers of fluid that are free to evaporate and heated from below a uniform convection pattern develops. This was observed by Bénard for the first time in 1900[15] and later explained by Block and Pearson.[16,17] The typical convection pattern that results is a honeycomb structure formed of uniformly spaced hexagonal figures. These cells are the result of gradients in the surface tension and a consequence of the Marangoni effect.[18] At this point there is a question that is quite a natural one to pose: how is it possible to recognize that what has been observed is Marangoni flow and not natural convection (buoyancy effect) as initially indicated by Rayleigh in his analysis?[19] In general, convection flows are always observed in liquid layers that are heated from the substrate; attributing the formation of Bénard cells to instability

triggered by the increase in density from the bottom to the top appears
therefore the simplest explanation. If convection is the cause of the flow,
the thickness of the evaporating layer should also affect the system.
Indeed, a critical thickness below which the liquid is stable should exist.
It has been experimentally observed, however, that Bénard cells form
even in thin films below the Rayleigh critical thickness, which means
that the origin of this phenomenon is on the surface and the surface
tension gradient is the driving force.

The spontaneous formation of cell-like patterns has been observed in
silica sol−gel films and are due to Bénard−Marangoni convections that
occur during solvent evaporation under stationary conditions.[20] The
characteristic hexagonal patterns originate from the gradient, which
arises from the difference in temperature between the substrate and the
liquid−air interface. If this difference is higher than the capability of the
fluid to conduct heat, a temperature gradient arises along the liquid film
or droplet. This means that the liquid film becomes unstable to thermal
fluctuation when the Marangoni number is higher than its critical value,
Ma_c. However, heating the substrate, as done by Bénard, can create this
temperature difference. The infinitesimal instabilities at the liquid-air
interface of an evaporating thin liquid layer (**Figure 2.6(a)**) can generate
some 'hot' and 'cool' regions that form local temperature gradients
(**Figure 2.6(b)**). The surface tension is higher in the regions of lower

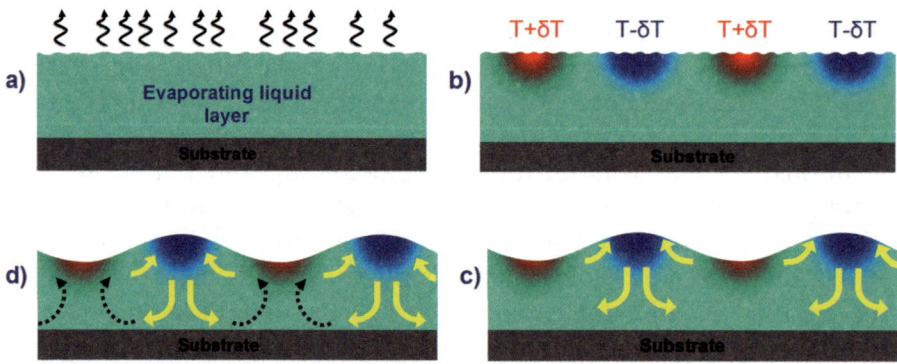

Figure 2.6   Infinitesimal instabilities at the liquid–air interface of an evaporating thin
liquid layer (a) can generate some 'hot' and 'cool' regions, which form
local temperature gradients (b). The surface tension is higher in the
regions of lower temperature and at the liquid−air interface; therefore,
the formation of hot and cold areas generates local surface gradients. This
situation creates a flow of liquid along the surface from the warmer
(lower γ) to the colder (higher γ) regions. In the colder area the liquid will
tend to accumulate and this creates a downward flow (c). Therefore, a
convective motion is generated that can be amplified by warming the
substrate (d).

temperature and at the liquid–air interface; therefore, the formation of hot and cold areas generates local surface gradients. This situation creates a flow of liquid along the surface from warmer (lower γ) to cooler (higher γ) regions. In the cold area the liquid will tend to accumulate and this creates a downward flow (**Figure 2.6(c)**). At the same time liquid will also flow from the warmer regions close to the substrate towards the warm surface areas to replenish the liquid that moved in to the cooler regions. Therefore, a convective motion is generated that can be amplified by warming the substrate (**Figure 2.6(d)**). A stationary state of this process is characterized by hexagonal patterns of convection, as observed by Bénard. Generally these convections are observed at high Marangoni numbers, which means that the viscosity and the thermal diffusivity should be low enough to allow the rise of the interfacial instabilities. In contrast, a higher thickness of the liquid film (see **eqn 2.1**) and a larger difference in temperature promotes the formation of Marangoni flows.

What now remains is to make a connection between the formation of the Bénard cells in the evaporating liquid films and the self-assembly of nanoparticles. If the system is optimized in terms of concentration, evaporation rate and viscosity, the nanoparticles should be able to accumulate on the substrate in correspondence of the borders of the Bénard cell, driven there by the Marangoni flow (**Figure 2.7(a)**). At the end of the process an interconnected structure of hexagons formed by well-packed nanoparticles should be observed. This is actually the case for different systems that have been experimentally observed such as gold nanoparticles in chloroform[21] and different types of nanocrystals.[22,23] An example of the hexagonal patterns formed upon evaporation of silver nanoparticle–hexane liquid layers is shown in **Figure 2.7(b)** and **(c)**. The technique has also been extended to zeolite silicalite nanoparticles that formed wrinkled honeycomb structures upon evaporation of the colloidal suspension.[24]

2.4 COMBINING COFFEE STAIN AND MARANGONI EFFECTS FOR THE SELF-ASSEMBLY OF NANOPARTICLES

In these first two chapters we have seen that evaporation is a complex, non-equilibrium phenomenon, and that capillary forces and surface tension play an important role in addressing the possibility of obtaining self-organized structures from colloidal solutions. The two effects can also interfere with each other. In an evaporating droplet, Marangoni vortexes can suppress the formation of the coffee-stain effect but can also be combined, at least in a sequential way, for making nanoarrays.

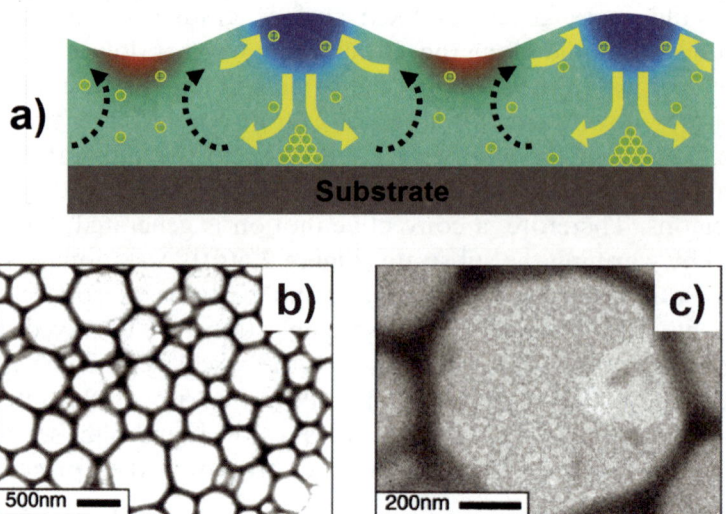

Figure 2.7 (a) The Bénard convections in an evaporating liquid film cause the formation of close-packed hexagonal structures by accumulating nanoparticles at the borders of the cells. (b) Hexagonal patterns formed upon evaporation of silver nanoparticles−hexane liquid layers. (c) A detail of a hexagon formed by the silver nanoparticles. (Figures b and c are adapted with permission from M. Maillard, *et al. J. Phys. Chem. B,* 2000, **104**, 11871. Copyright 2000, American Chemical Society.)

An interesting example of this combination has been shown by Cai *et al.*[25] where self-assembled 100 nm polystyrene (PS) nanoparticles in 2D hexagonal (**Figure 2.8(a)**) and 1D strip like patterns (**Figure 2.8(b)**) were produced using the evaporating Marangoni flow. The formation of these patterns is the result of a controlled evaporation of a suspension of ethanol−PS nanoparticles which is spread as thin films on two different substrates with contact angles of 32 and 0°, respectively. If we look very carefully at the image in **Figure 2.8(a)** it appears that the ordered 2D structure of fluorescent dots is in reality formed by an array of rings. How is it possible to get such peculiar structures of nanoparticles rings? Clearly we can suspect that the coffee-stain effect should also have appeared at some stage of the evaporation process. Let's see the details of the experiment to understand how it is possible to get such result. The evaporation process is conducted in an atmosphere of wet N_2, which is a very important factor because it allows for the condensing of water near the contact line of the receding ethanol film. The concentration of ethanol near the contact line is lower than the inner volume and this triggers the rise of the concentration

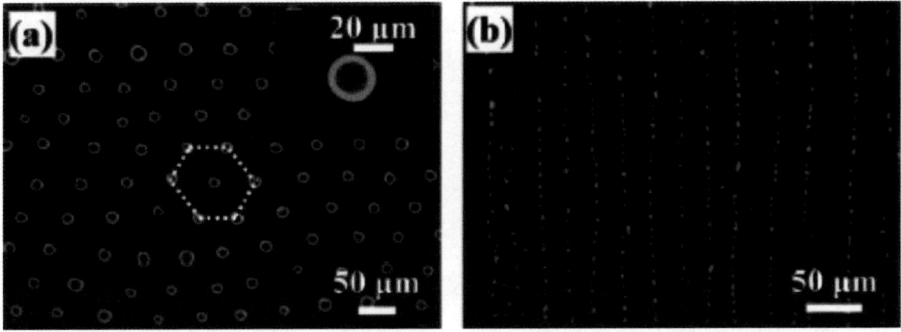

Figure 2.8 Images taken by fluorescence microscopy of the self-assembled patterns formed by carboxylated polystyrene nanoparticles. (a) Hexagonal regular arrays of rings (inset in the figure) are formed on a silica substrate (contact angle $\sim 32°$). (b) Regular strip-like patterns of nanoparticles formed on silica substrate of higher hydrophilicity (contact angle $\sim 0°$). (Reprinted with permission Y. Cai *et al.*, *J. Am. Chem. Soc.*, 2008, **130**, 6076. Copyright 2008, American Chemical Society.)

gradient between the contact line and the rest of the liquid film. The difference in the surface tension between the water (72 mN m^{-1}) and the ethanol (22 mN m^{-1}) generates a surface tension gradient, which in turn produces a Marangoni flow of ethanol to carry the nanoparticles in the direction of the contact line into the condensing water (**Figure 2.9**). Ethanol evaporates much faster than water and this means that the receding evaporating film of ethanol, in the case of a non-wettable substrate, leaves behind water droplets containing the nanoparticles. It is easy at this stage to imagine that after the evaporation of water, because of the coffee-stain effect, what will actually form is a ring of nanoparticles. However, what remains to be explained is why such an ordered array is observed? We have to introduce another type of instability, which will be described in detail in the next chapter: the 'fingering instability', a phenomenon that occurs at the interface between two fluids of different densities and that also involves, in several cases, a Marangoni flow.[5] **Figure 2.10** describes the details of the process as explained by Cai. At first, water condensation at the contact line results in the formation of small fingers at regular distances that are likely triggered by the fluctuation of the receding contact line. The fingers eventually grow and detach from the interface, while in the middle of two large and equivalent neighboring fingers a new smaller finger nucleates and results in alternate big and small fingers on the contact line. The receding film of ethanol will leave behind, therefore, droplets of water in alternate and regularly spaced positions due to the mechanism of finger formation.

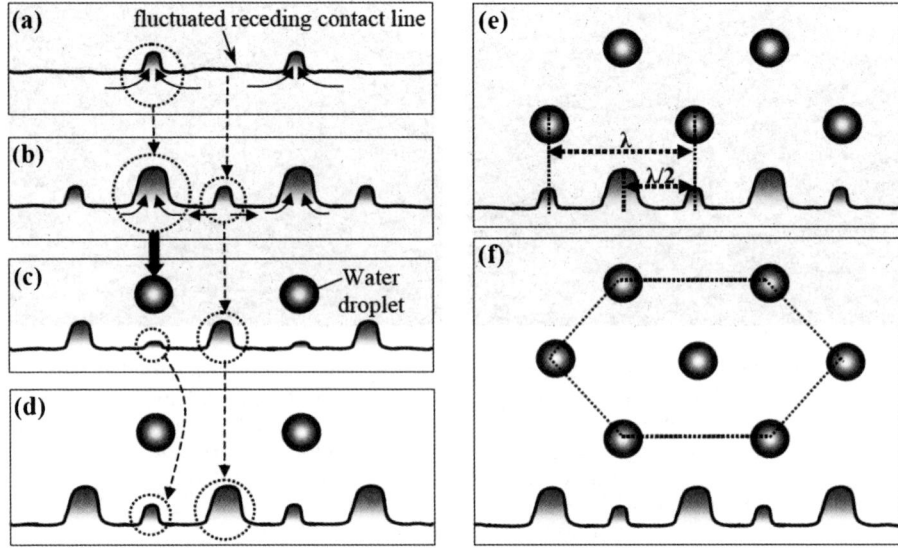

Figure 2.9 The formation of hexagonal water droplets by evaporation of a nanoparticles–ethanol suspension spread on a slightly non-wettable substrate (contact angle 32°C). (a) Water fingers start to form from condensation of the wet N_2 stream. The curved arrows indicate the local flow of liquid, which contains the nanoparticles, near the finger toward the propagating finger front. (b) Growth of the fingers and initiation of new smaller fingers between two neighboring fingers. (c) Detachment of fully grown fingers from the contact line to form water droplets and growth of smaller fingers in (b). (d) Further growth of fingers before detachment. (e) and (f) The steps shown from (a) to (d) are repeated to form hexagonal water droplet arrays. (Adapted with permission Y. Cai *et al., J. Am. Chem. Soc.*, 2008, **130**, 6076. Copyright 2008, American Chemical Society.)

The substrate in evaporation phenomena is really important.[26] It is part of the game and can be successfully used to design the organization of nanostructures. Repeating the experiment using a wettable substrate, in fact, drastically changes the results and linear stripes of nanoparticles are formed upon evaporation. The condensed water will completely wet the substrate and produces a very thin layer of water. At this point the large difference in surface tension and the fingering instability push the growth of fingers into ordered dot stripes.

However, this method seems flexible enough that it can be adapted to produce porous polymer films in a kind of replica process.[27] After the formation of regular arrays of water droplets on a substrate they are dip-coated in a $PS-CHCl_3$ solution, which upon evaporation leaves porous PS thin films whose porosity reproduces exactly the regular array of the water droplets.

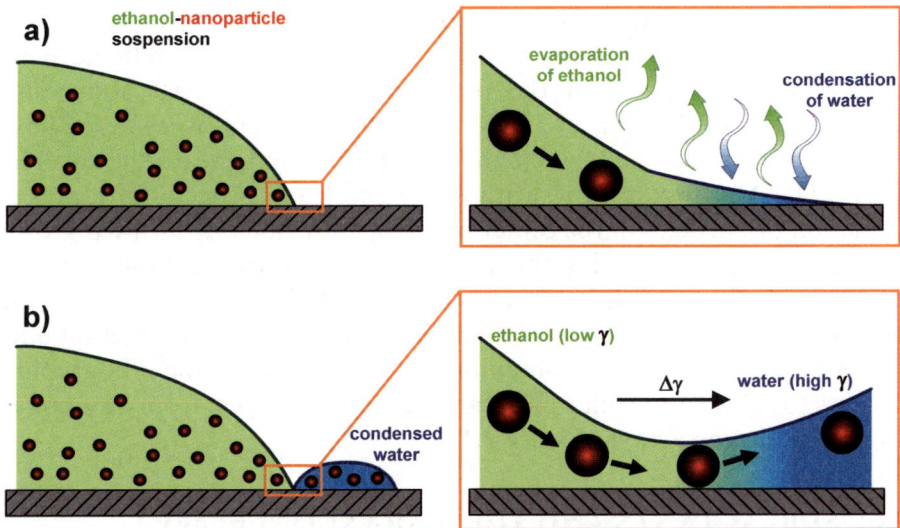

Figure 2.10 Marangoni flow-induced self-assembly of nanoparticles. (a) The ethanol2nanoparticles suspension is exposed to a wet N_2 stream to induce the evaporation of ethanol and the condensation of water near the three-phase contact line and a concentration gradient near the contact line arises. (b) The concentration gradient and the difference between the surface tensions of ethanol (γ_{EtOH} = 22 mN m^{-1}) and water (γ_{H2O}) 72 mN m^{-1}) create a surface tension gradient, which drives the nanoparticles2water toward the direction of the receding contact line. (Adapted with permission Y. Cai *et al., J. Am. Chem. Soc.*, 2008, **130**, 6076. Copyright 2008, American Chemical Society.)

2.5 LITHOGRAPHY USING THE MARANGONI EFFECT

Marangoni flows, as we have seen, induce recirculation of particles within an evaporating colloidal system and can suppress the coffee-stain effect or can form specific self-ordered, patterned, structures. Is it possible to really control the Marangoni effect and use it as a lithographic patterning tool? The answer relies on the capability of inducing Marangoni flows within spatially controlled evaporating systems. Harris and Lewis[28] proposed a simple and elegant solution that comprised of a dried non-aqueous colloidal film beneath a patterned mask. The mask produces periodic variations between the free and the hindered regions of the evaporating films. In correspondence of the open areas of the mask, cooler regions are formed on the film surface. This gives rise to periodic temperature gradients, which also means surface tension gradients, which in turn generate Marangoni flows within areas of the films that follow the design of the mask. The recirculating flow within these cells promotes the deposition of the particles to form regular structures (**Figure 2.11**), as we have previously seen. In particular, in hexagonal recirculating flow, the cells develop almost immediately when the patterned mask is placed above the evaporating colloidal layer. Each one of these cells is centered just below a hole in the mask, while the interfaces form beneath the covered areas. The Marangoni flow, with the evaporation process, produces a depletion of particles in the center of the open regions whose area grows in size with drying. During this process the particles accumulate in the non-evaporating regions to form a pattern on the substrate that is the negative of the drying mask. Tuning the colloidal volume fraction and the mask design allows for obtaining a good control of the lithographic process to fabricate regular patterns of different shapes, from hexagonal to parallel linear arrays (**Figure 2.12**).

2.6   SURFACTANTS AND THE MARANGONI EFFECT

One of the consequences of the rise of the Marangoni effect in an evaporating droplet is the competition with the outward directed flow

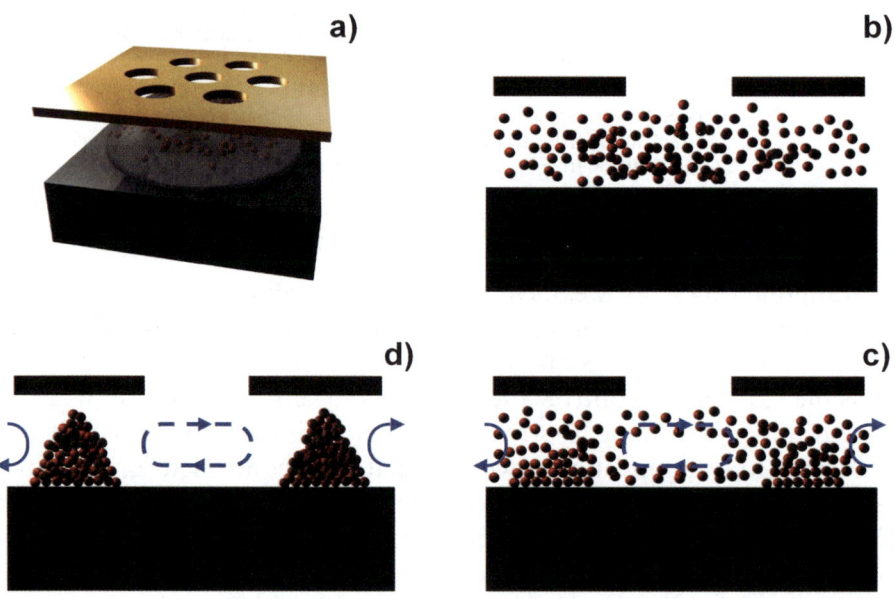

Figure 2.11   Lithographic technique that employs controlled Marangoni flows to produce patterns of different types. (a) A mask is applied on the surface of an evaporating colloidal non-aqueous solution (b). Convective cells (c) accumulate the particles (d) in correspondence of the non-evaporative regions (masked areas). (Adapted with permission D. J. Harris *et al. Langmuir*, 2008, **24**, 3681. Copyright 2008, American Chemical Society.)

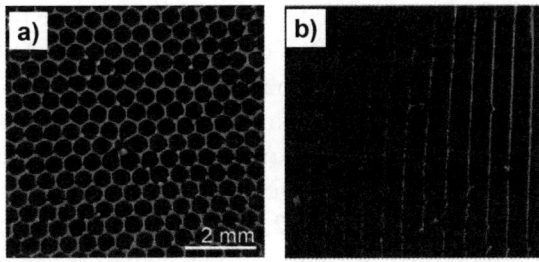

Figure 2.12 Optical images of a hexagonal (a) and a linear pattern (b) obtained by evaporation of a silica colloidal solution under patterned masks. (Reprinted with permission D. J. Harris *et al. Langmuir*, 2008, **24**, 3681. Copyright 2008, American Chemical Society.)

due to the coffee-stain effect with complete or partial suppression of the ring formation at the contact line.[29] Both theory and experiments have shown that even a small concentration of insoluble surfactants on the liquid–air interface, as low as 300 molecules per μm^2, is able to reduce in a significant way the radial flow produced by the Marangoni effect.[30] This is the basic reason, as we have shown in Chapter **1**, why in solvents, which are easily contaminated, such as water, the Marangoni flow is quite hard to observe. We have also noticed, however, that the Marangoni effect may counterbalance the flow that rises during the coffee-stain effect and can suppress the ring formation leaving a homogeneous halo at the end of droplet evaporation. This happens without the addition of any surfactant.

In general, the Marangoni flow is proportional to the surface tension gradient, c. Indeed if the gradient reaches a threshold value close to the contact line, an inward flow is generated and it compensates the outward flow. An example that makes this point clearer is the evaporation of a droplet containing a dissolved polymer and a surfactant. The coffee-stain effect produces an accumulation of surfactant close to the contact line which on turns produces a local decrease in surface tension and a surfactant-driven Marangoni flow.[31] The result is a levelling effect caused by the accumulation of surfactant at the droplet edge and the formation of a flat polymeric surface.

This effect is also observed in colloidal solutions such as in the case reported by Still and co-workers for an aqueous suspension of polystyrene spheres containing sodium dodecyl sulfate as the surfactant.[32] This example is particularly interesting because it shows how the surfactant may have so much influence on the motion of the particles, which becomes quite hectic. As soon as the contact line is pinned, the convective flow, which is triggered by the coffee-stain effect, concentrates the surfactant with consequent reduction, such as in the previous case, of the water−air surface tension. This creates an inward Marangoni flow, which is strong enough to push the particles toward the drop centre. However, the movement of these particles is not over. As soon as they leave the droplet edge the Marangoni effect becomes weaker and they again are pushed toward the contact line by the incoming coffee-stain flow, which is illustrated in **Figure 2.13**. This movement of the particles creates what can be called an 'eddy' zone, which is the corona region where the particles circulate back and forth between the contact line of accumulation and a depletion zone (**Figure 2.14**).

This is not the only case that we have encountered particles, which return back from the contact line during the evaporation of a droplet. In Chapter **1** we have already observed that under some conditions a capillary lateral force can repel the particles back from the edge.[33] However, they do not remain entrapped such as in the presence of surfactant into a swirling zone and simply tend to form a homogeneous layer.

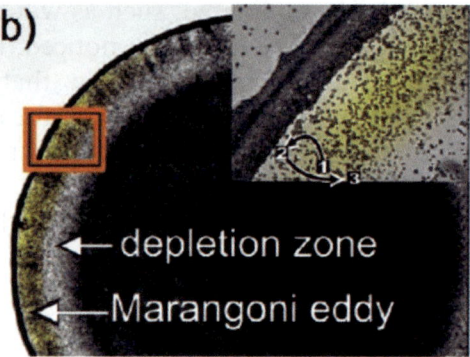

Figure 2.13   Snapshot during the evaporation of a water drop containing 0.5 wt% PS particles (d = 1330 nm) and 0.5 wt% SDS at magnifications of 5× (large) and 63× (inset). The Marangoni eddy is highlighted yellow where particles from the bulk flow toward the edge but are repelled on the surface toward the centre of the drop. In the 63× picture, the motion of an exemplary single sphere is highlighted by numbers 1−3, respectively, at the time of the picture, 0.25 s later (*i.e.* sphere next to the edge), and 0.5 s later (*i.e.* sphere repelled). Two straight lines connecting the numbers essentially give its trajectory. (Reproduced with permission T. Still *et al.*, *Langmuir*, 2012, **28**, 4984. Copyright 2012, American Chemical Society.)

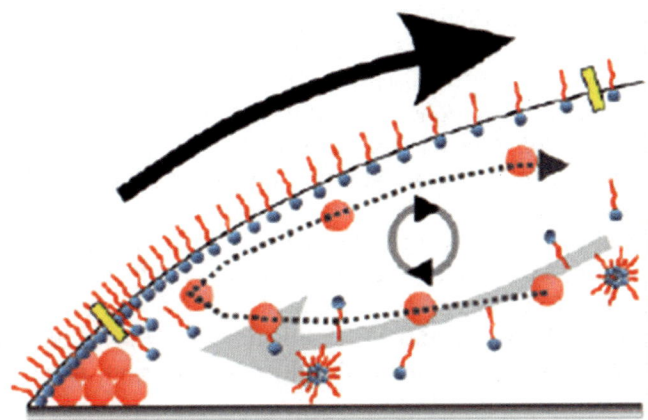

Figure 2.14 A schematic description of the Marangoni eddy. Surfactant molecules from the bulk are pushed to the pinned contact line, where they try to reach the water−air interface. The locally increased surface concentration of sodium dodecyl sulfate decreases the local surface tension and leads to a surface Marangoni flow toward the center of the drop. The surfactant can be present as a single molecule or micelle. (Reproduced with permission T. Still *et al.*, *Langmuir*, 2012, **28**, 4984. Copyright 2012, American Chemical Society.)

REFERENCES

1. X. Xu and J. Lo, *Appl. Phys. Lett.*, 2007, **91**, 124102.
2. C. Marangoni, *Nuovo Cimento Ser. 2*, 1872, **7–8**, 301.
3. Carlo Marangoni (1840–1925) gave the first description of the effect in his dissertation thesis (1865) at the University of Pavia "Sull'espansione delle gocce liquide (About expansion of liquid droplets.)".
4. X. Fanton, A. M. Cazabat and D. Quéré, *Langmuir*, 1996, **12**, 5875.
5. A. M. Cazabat, F. Heslot, S.M. Troian and P. Carles, *Nature*, 1990, **346**, 824.
6. R. Tadmor, *J. Colloid. Interface Sci.*, 2009, **332**, 451.
7. D. J. Taylor and D. P. Birnie III, *Chem. Mater.*, 2002, **14**, 1488.
8. D. P. Birnie III, *J. Mater. Res.*, 2001, **16**, 1145.
9. H. Kozuka and M. Hirano, *J. Sol−Gel Sci. Technol.*, 2000, **19**, 501.
10. H. Kozuka, Y. Ishikawa and N. Ashibe, *J. Sol−Gel Sci. Technol.*, 2004, **31**, 245.
11. D. P. Birnie III, D. M. Kaz and D. J. Taylor, *J. Sol−Gel Sci. Technol.*, 2009, **49**, 233.
12. D. E. Haas and D. P. Birnie III, *J. Mater. Sci.*, 2002, **37**, 2109.
13. K.-H. Wu, S.-Y. Lu and H.-L. Chen, *Langmuir*, 2006, **22**, 8029.
14. S.-Y. Lu, H.-L. Chen, K.-H. Wu and Y.-Y. Chen, *Langmuir*, 2007, **23**, 10069.
15. H. Bénard, *Rev. Gén. Sci. Pures Appl. Bull Assoc. Franc. Avan. Sci.*, 1900, **11**, 1261.

16. M. Block, *Nature*, 1956, **178**, 650.
17. J. R. A. Pearson, *J. Fluid Mech.*, 1958, **4**, 489.
18. M. F. Schatz, S. J. Van Hook, W. D. McCormick, J. B. Swift and H. L. Swinney, *Phys. Rev. Lett.*, 1995, **75**, 1938.
19. L. Raleygh, *Phil. Mag.*, 1916, **32**, 529.
20. H. Uchiyama, Y. Miki, Y. Mantani and H. Kozuka, *J. Phys. Chem. C*, 2012, **116**, 939.
21. C. Stowell and B. A. Korgel, *Nano Lett.*, 2001, **1**, 595.
22. M. Maillard, L. Motte, A. T. Ngo and M. P. Pileni, *J. Phys. Chem. B*, 2000, **104**, 11871.
23. M. Maillard, L. Motte and M. P. Pileni, *Adv. Mater.*, 2001, **12**, 200.
24. H. Wang, Z. Wang, L. Huang, A. Mitra and Y. Yan, *Langmuir*, 2001, **17**, 2572.
25. Y. Cai and B.-M. Z. Newby, *J. Am. Chem. Soc.*, 2008, **130**, 6076.
26. W. D. Ristenpart, P. G. Kim, C. Domingues, J. Wan and H. A. Stone, *Phys. Rev. Lett.*, 2007, **99**, 234502-1.
27. Y. Cai and B.-M. Z. Newby, *Langmuir*, 2009, **25**, 7638.
28. D. J. Harris and J. A. Lewis, *Langmuir*, 2008, **24**, 3681.
29. H. Hu and R. G. Larson, *J. Phys. Chem. B*, 2006, **110**, 7090.
30. H. Hu and R. G. Larson, *Langmuir*, 2005, **21**, 3972.
31. T. Kajiya, W. Kobayashi, T. Okuzono and M. Doi, *J. Phys. Chem. B*, 2009, **113**, 15460.
32. T. Still, P. J. Yunker and A. G. Yodh, *Langmuir*, 2012, **28**, 4984.
33. B. M. Weon and J. H. Je, *Phys. Rev. E*, 2010, **82**, 015305, (R).

CHAPTER 3

The Lord of the Rings: Stick and Slip Motions and Self-Assembly During Coffee-Stain Formation

The content of this chapter is strictly related to the previous two subjects. We have in fact just discussed how an evaporating colloidal liquid could show such surprising features that drive the system directly into self-assembly and self-organization. A droplet of a colloidal solution can leave well shaped rings after drying while Marangoni flows in the same droplets may even lead to self-ordered complex geometrical structures. However, please do not forget that this is very much what happens in an ideal world with very well or almost perfectly controlled experimental conditions. In the majority of the cases we have to face deviations from the ideal response and we should also be very much aware that several other chemical−physical phenomena can occur to make the situations much more complex.

3.1 PINNING: DEPINNING IN AN EVAPORATING DROPLET

In Chapter 1 we saw that one condition for the formation of rings in an evaporating colloidal droplet is pinning, *i.e.* the blockage of the droplet contact line, which does not recede during evaporation. The accumulation of particles at the droplet edge also contributes to a self-pinning effect.[1] Pinning phenomena are due to chemical heterogeneities of the solid substrate or the presence of surface defects. Evaporation is, however, a dynamic and not an equilibrium process and instability phenomena where depinning of the contact line can occur. Depinning during evaporation of a colloidal droplet can be observed several times before the evaporation has gone to completion. This produces strange

Water Droplets to Nanotechnology: A Journey Through Self-Assembly
By Plinio Innocenzi, Luca Malfatti and Paolo Falcaro
© P. Innocenzi, L. Malfatti and P. Falcaro 2013
Published by the Royal Society of Chemistry, www.rsc.org

and irregular concentric patterns at the end of the process. We can even state that if the coffee-stain effect is not properly managed to achieve well-defined rings that this can be the most common finding. This is a typical *stick and slip* motion As we will see later in this chapter, if controlled it can become another tool for transforming self-assembled nanoparticles into well defined and specific patterns.

To understand how depinning gives rise to concentric and irregular rings we can start by analyzing the case of a small droplet on an ideal flat and smooth surface; the contact angle at equilibrium, θ_0, is given by the well known Young's equation:

$$\gamma_{SV} - \gamma_{SL} = \gamma_{LV} \cos \theta_0 \tag{3.1}$$

where γ_{SV}, γ_{SL} and γ_{LV} are the surface tensions at the solid−vapor, solid−liquid and liquid−vapor interfaces, respectively. To describe the dynamic of the three phases (solid, liquid, gas) at the contact line in a drying drop of a pure liquid (such as ethanol and water) the simple model elaborated by Shanahan and co-workers[2,3] is well in agreement with the experimental findings. The progress of evaporation, still assuming the pinning of the contact line, causes a slight shrinkage of the droplet with a concurrent small decrease of the contact angle ($\theta_0 - \delta\theta$) (**Figure 3.1**). The Young equation, at equilibrium, indicates that there are no forces acting at the contact line ($\gamma_{SV} - \gamma_{SL} - \gamma_{LV} \cos \theta_0 = 0$) but with a small reduction of the contact angle, a force, $\delta \vec{F}$, which pushes toward the center of the evaporating drop arises:

$$\delta \vec{F} = \gamma_{LV} \cos (\theta_0 - \delta\theta) - \gamma_{LV} \cos \theta_0 \approx \gamma_{LV} \sin \theta_0 \delta\theta \tag{3.2}$$

The possibility that the triple contact line, air−liquid−solid, will move depends, however, on the intensity of the force that should be high enough to allow crossing an intrinsic energy barrier, U, which prevents depinning of the contact line. The differential of this energy, $\partial U/\partial r$, is a force that locally opposes the movement of the triple contact line given by $\delta \vec{F}$. Equilibrium between the two forces is reached just at the threshold of depinning:

$$\frac{\partial U}{\partial r} = \delta \vec{F} = \gamma_{LV} \sin \theta_0 \delta\theta \tag{3.3}$$

however, if the contact angle decreases slightly more, $\delta \vec{F}$ could become large enough to cross the energetic barrier, which means that the contact line will

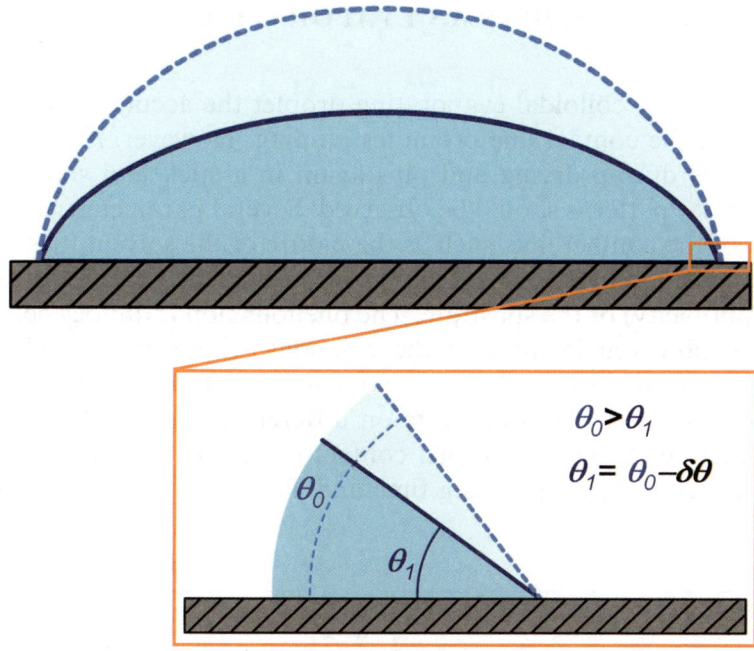

Figure 3.1 The pinning of the contact line causes a slightly shrinkage of the droplet with a concurrent small decrease of the contact angle ($\theta_0 - \delta\theta$) (enlargement of the contact line region at the bottom).

recede and the droplet is depinned. The change of the contact angle from the equilibrium depends very much on the system, such as the type of surface and liquid used.[3] This is a simple representation of a pinning–depinning process in an evaporating droplet of a pure liquid. We can also change perspective and describe this as a stick and slip process. The pinning is the sticking stage during evaporation whilst the recession of the triple contact line represents the slipping. On the other hand, this pinning–depinning or stick and slip process during drying of the liquid droplet can be observed several times under the proper conditions, at least until the receding droplet is able to pin again before full evaporation.

There is another point that also needs to be underlined which is the role of the substrate, in particular the hydrophobicity. Experiments done on an evaporating droplet of ethanol on substrates of different hydrophobicity have shown, in fact, the dependence of droplet lifetime on contact angle.[4] The evaporation of the droplet on a surface of high contact angle gives an almost regular recession of the triple line during liquid evaporation maintaining a constant θ, while at low contact angles θ decreases with time which gives rise to pinning. The evaporation time is longer in the pinned droplets.

3.2   STICK AND SLIP IN AN EVAPORATING COLLOIDAL DROP

In the case of a colloidal evaporating droplet the accumulation of the particles at the contact line promotes pinning. However, if the contact line recedes during drying and pins again in a stick and slip motion, some regular patterns should be observed. Several parameters affect this motion of the contact line, such as the nature of the solvent, the type of particles, their dimension and composition and the surface properties (hydrophobicity) of the substrate. The phenomenon is quite general and could be observed in most of the colloidal evaporating droplets.[5] A good example to show is the case of a water$-$TiO$_2$ nanoparticles (NPs) droplet that was left to evaporate on different types of substrates.[6] In **Figure 3.2** the changes in droplet contact radius, R, and contact angle, θ, as a function of evaporating time and TiO$_2$ NPs concentrations are

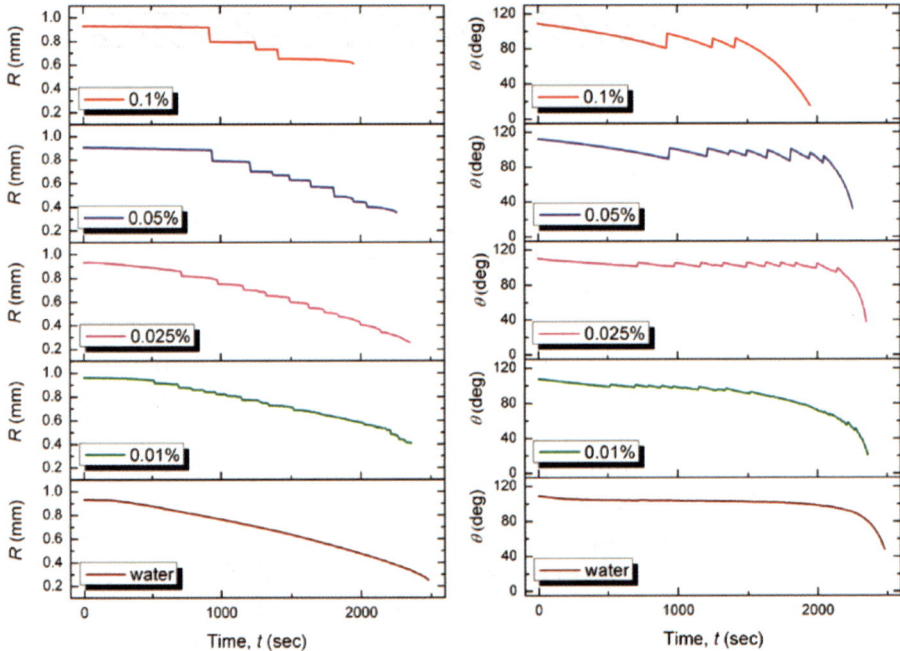

Figure 3.2   Evolution of the contact radius, R (mm), with time, t (s) (left column) and contact angle, θ (°), with time, t (s) (right column), during evaporation on Cytop ($\theta_0 \approx 110°$) of a water droplet containing different TiO$_2$ nanoparticles concentrations. The change of R and θ as a function of evaporation time of a pure water droplet are shown in the first graph from the bottom on both the columns. (Reprinted with permission from D. Orejon *et al.*, *Langmuir*, 2011, **27**, 12834. Copyright 2011, American Chemical Society.)

shown. The substrate is the hydrophobic Cytop ($\theta_0 \approx 110°$).[3] In the case of a pure water droplet under the same conditions of evaporation there is no pinning of the contact line onto such a hydrophobic surface. The contact line in the water−TiO$_2$ NPs droplet, instead, is not only pinned but 'jumps' several times in correspondence of a change of contact angle in a stick and slip process that lasts until evaporation has finished. The concentration of TiO$_2$ NPs also has an important role because the stick and slip behavior was enhanced with the increase of NPs concentration and, at the same time, the number and distance of the jumps depend on the NPs content, as can be seen in **Figure 3.2**. After every jump when the contact line pins again, the NPs accumulate (sticking phase) and after each depinning step (slip motion) an irregularly concentric pattern is left behind (**Figure 3.3**). The rings correspond to the stick period and the clean areas correspond to the

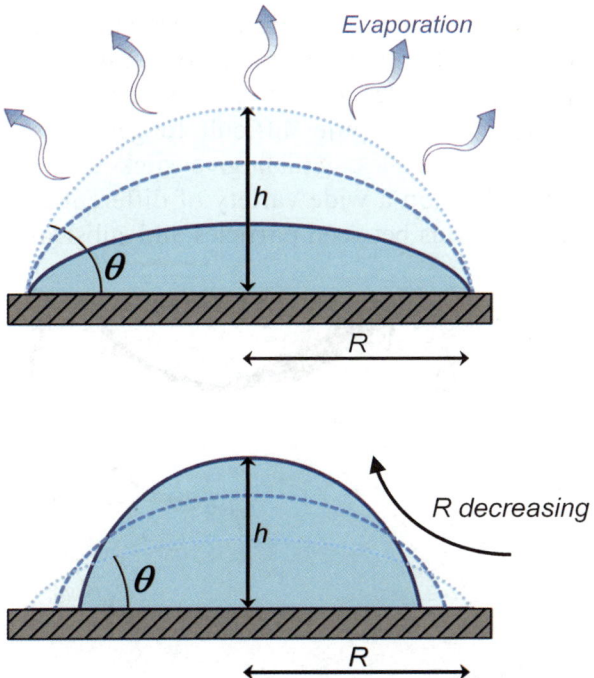

Figure 3.3   (a) Representation of a pinned droplet. Radius is fixed, both contact angle and drop height decrease in time. (b) Representation of triple line jump. Excess free energy increases until it exceeds the hysteretic energy barrier. At this point the drop depins and the contact radius, R is reduced, while the contact angle, θ, and drop height, h, are both increased. (Adapted with permission from J. R. Moffat *et al.*, *J. Phys. Chem. B*, 2009, **113**, 8860. Copyright 2009, American Chemical Society.)

slips. In general, the accumulation of nanoparticles at the contact line increases the depinning force and longer pinning is reflected in a greater jump of the contact line.

If we observe **Figure 3.2** more carefully we should be able to notice, however, quite an interesting variation of R and θ curves with evaporation time. The radius of the droplet after every jump, due to depinning, shows a steady decreasing step and then a continuous decrease until the next jump. The contact angle appears to be similar but the step is up and not down. To understand how this happens let's have a look again at the evaporation droplet profile. In the case of the pinned contact line, the radius R is fixed while both the contact angle and the droplet height decreases with time. This picture changes when the triple contact line is depinned; in fact R decreases but the contact angle and the droplet height increases (**Figure 3.4**).[6]

We have just seen what can happen in a colloidal droplet that is free to evaporate on a solid substrate. The result, because of stick and slip motions[7], can easily be the formation of irregular patterns.[8] These concentric patterns are basically the result of a random process which sticks the evaporating droplet in some points of the substrate without any spatial regularity. It is quite difficult to generalize the cause of pinning and depinning cycles, which give stick and slip motion, it depends at the very end on a wide variety of different parameters such as the mutual interactions between particles and substrate. A good way

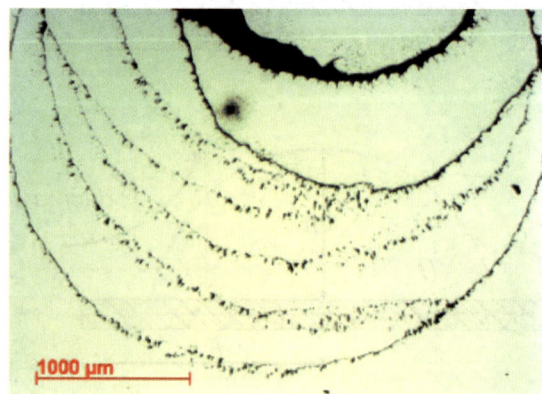

Figure 3.4   Image showing the stick and slip rings formed upon evaporation of an ethanol droplet containing a 0.1% by weight suspension of TiO_2 nanoparticles. Rings of accumulated particles correspond to 'stick' in the stick−slip cycle. (Reprinted with permission from J. R. Moffat *et al.*, *J. Phys. Chem. B*, 2009, **113**, 8860. Copyright 2009, American Chemical Society.)

of explaining stick and slip, however, rather than using very precise physical–chemical models[7,9] is to view the process essentially as stochastic. The stick–slip motion of a water droplet containing polystyrene particles of 0.88 μm produces an irregular series of jumps of the contact line[10] as shown in **Figure 3.5**. The distance between successive stick points of the contact line and the moving and pinning times are irregular and exactly as seen in the previous case of TiO_2 NPs. The distances such as the elapsed time between two depinning events have also a very broad distribution (insets in the **Figure 3.5**). We should consider that not necessarily all the particles flow to the contact line and some of them can randomly pin in some inner points of the liquid–substrate interface and with the receding of the droplet upon depinning they can become a new anchoring site for the formation of the successive pinning line[10] (**Figure 3.6**). This explains the random formation of irregular concentric rings.

The formation of multiple concentric rings upon evaporation of a colloidal droplet is not limited to the case of particles. The multi-rings formed by drying of drops containing different concentrations of DNA are another interesting example[11]. The capability of pattern formation

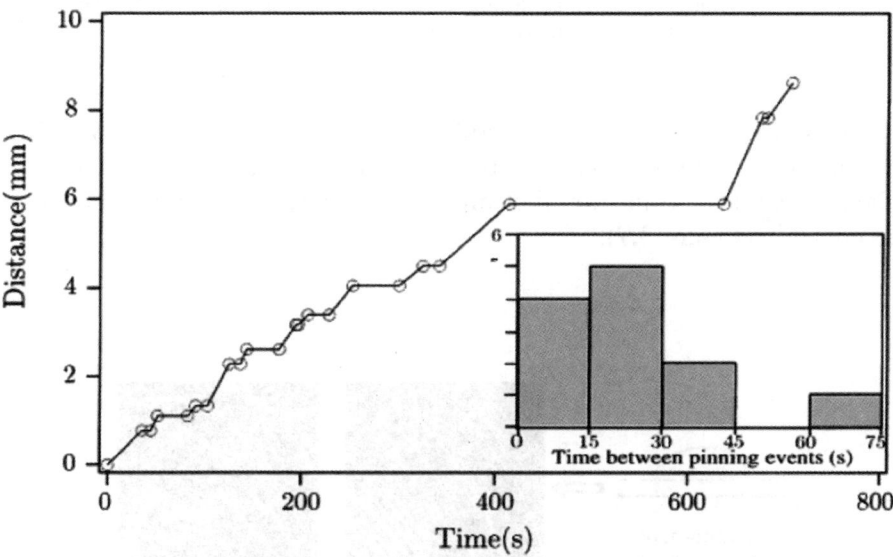

Figure 3.5   Stick–slip motion of the contact line of a water-drying droplet containing 0.88 μm diameter polystyrene particles. The distance on the *y*-axis shows the moving steps of the contact line from the drop edge. A bar graph of the distribution of times for each moving step is shown in the inset. (Reprinted with permission from L. Shmuylovich *et al.*, *Langmuir*, 2002, **18**, 3441. Copyright 2002, American Chemical Society.)

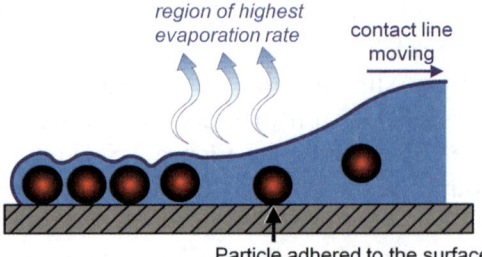

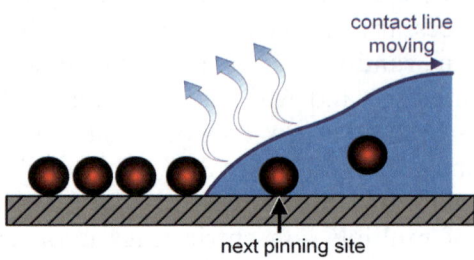

Figure 3.6 Not all the particles flow to the contact line, some of them can randomly pin in some inner points of the liquid−substrate interface (a) and with the receding of the droplet upon depinning they can become a new anchoring site for the formation of the successive pinning line (b). (Adapted with permission from L. Shmuylovich *et al.*, *Langmuir*, 2002, **18**, 3441. Copyright 2002, American Chemical Society.)

remains intact in a binary colloidal suspension of DNA−silica nanoparticles but in this case the composition of the binary system plays a critical role for the multi-ring appearance and the final geometry[12] (**Figure 3.7**).

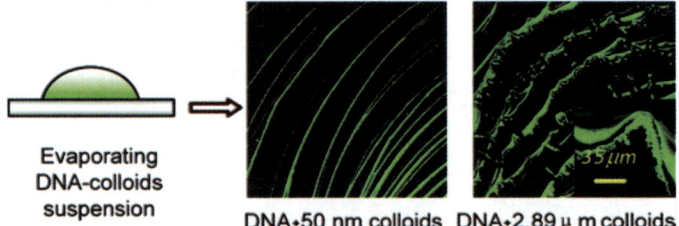

Figure 3.7 Multiple-ring stains inside the DNA droplet added with carboxyl end-functionalized silica colloidal particles of varied particle sizes 50 nm and 2.89 μm. (Reprinted with permission from L. Zhang *et al.*, *Langmuir*, 2008, **24**, 3911. Copyright 2008, American Chemical Society.)

3.3 THE LORD OF THE RINGS, CONTROLLING THE STICK AND SLIP MOTION

It appears that drying and stick and slip motions should be more strictly controlled for achieving real ordered structures. An elegant solution is to confine the geometry of the evaporating droplet by using a spherical lens to control the contact line; this allows for obtaining concentric rings of extremely well-defined geometry.[13–15] This method has been successfully applied for producing ordered structures in organic polymeric films[16–18] but it has also been demonstrated to be a very feasible technique to be applied to a colloidal solution, which is the case of our interest.

The method, in general, is based on constraining an evaporating droplet on a flat surface (silicon, mica, ITO) using an axially symmetrical spherical lens (silica glass) which is put in contact with the substrate; this produces a convex meniscus interface (capillary bridge) at the edge of the droplet with a rise of capillary forces (**Figure 3.8**).[19] The geometrical restriction constrains the evaporation at the capillary edge where it is faster. At this point of the evaporation process the coffee-stain effect starts to come into play. The particles, or molecules, accumulate at the edge driven by the capillary flow and self-pin the contact line (stick). The initial contact angle of the meniscus, θ_i, gradually decreases with the evaporation of the solvent until a critical angle, θ_c, is reached. At this angle the depinning force, as previously described, becomes large enough to push the contact line forward in the direction of the center of the droplet (slip). This produces a repetitive movement of the contact line in a controlled stick−slip process from the edge of the droplet to the center of the sphere−surface contact area.

Extremely well defined and regular concentric rings, which cover all the area of the droplet with the exception of the lens contact region, are finally formed. The rings exhibit a gradient concentric structure, which means that the size, height and interdistance of the rings slowly decrease from the outer part moving to the center. These parameters can be designed, within some limits, by controlling the concentration of the colloidal solution, the solvent and the substrate.

However, this technique does not restrict the type of geometries that can be obtained to concentric rings. The shape of the constraining tool very much affects the pattern's shape. A square pyramid surface produces a gradient of concentric square stripes from the capillary edge to the center,[20] whilst a triangular slice sphere allows obtaining more complex geometries with triangular symmetries[20] (**Figure 3.9**). This technique, which has been developed initially for patterning polymers,

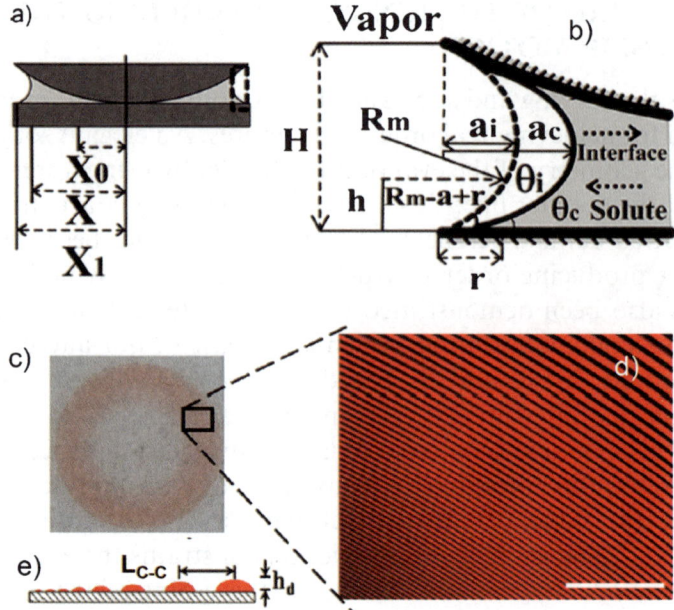

Figure 3.8   Concentric rings obtained by a toluene solution containing a linear
conjugated polymer, poly[2-methoxy-5-(2-ethylhexyloxy)-1,4-phenylene-
vinylene] (MEH-PPV) as non-volatile solute. (a) Schematic cross-section
of a sphere-on-flat configuration; the MEH-PPV solution is poured on
the substrate and the concave sphere produces a geometrical constraint
to evaporation. X_1, X, and X_0 are the radii of outermost, intermediate,
and innermost rings from the center of sphere−substrate contact,
respectively. (b) Detail of the capillary edge which is formed by the liquid
between the sphere and the substrate. The initial contact angle of the
meniscus, θ_i, gradually decreases with the evaporation of the solvent until
a critical angle, θ_c, is reached. (c) The digital image of entire gradient
concentric ring patterns formed by the deposition of the solute in the
geometry in (a). (d) An enlargement of the fluorescence image of MEH-
PPV ring patterns. The scale bar is 200 μm. (e) The rings become smaller
and the height decreases as the solution front moves inward. (Reprinted
with permission from J. Xu *et al.*, *Phys. Rev. Lett.*, 2006, **96**, 066104.
Copyright 2006, American Physical Society.)

has also been very effective as a replica method for obtaining concentric
metallic rings such as gold.[21]

3.4   RINGS FROM COLLOIDAL SOLUTIONS

Evaporation in a confined environment is not restricted to organic
polymers and it works also very well in the case of colloidal solutions;
this means that nanoparticles and nanostructures self-organize to form
concentric ordered structures driven by capillary forces, which rise in
the restricted region of evaporation. There are some examples of

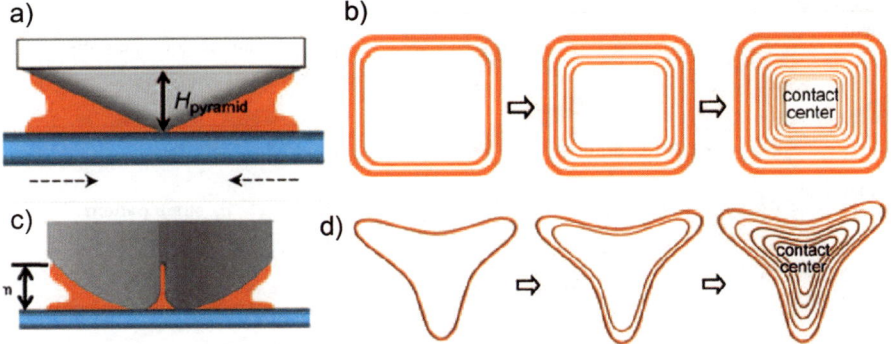

Figure 3.9 (a) A square pyramid is placed on a Si substrate to hold the solution; this particular shape produces gradient concentric squares. (b) Top schematic view of the formation of gradient concentric square stripe propagating from the capillary edge toward the pyramid/Si contact center during solvent evaporation. (c) Triangular-slice-sphere-on-Si geometry and (d) the concentric triangular-like concentric pattern which are formed upon evaporation. (Reprinted with permission from S.W. Hong *et al.*, *Angew. Chem., Int. Ed.*, 2009, **48**, 512. Copyright 2009, John Wiley & Sons.)

particular interest. The first one is the self-organization of graphene,[22] in particular, the reduced form of graphene oxide (RG-O) which is similar to pure graphene but contains residual oxygen groups and structural defects.[23] The importance of the interaction between the substrate and particles is particularly clear in this case. In fact, if the experiment of self-organization is done just by employing an aqueous suspension of RG-O only a chaotic structure is observed. After evaporation, however, order arises if a poly(ionic liquid) (PIL) is added. PIL is not only a stabilizer, it helps to obtain a homogeneous RG-O solution. If the deposition is carried out on a negatively-charged substrate the poly(ionic liquid) positive charge drives self-organization by promoting the pinning process *via* electrostatic interaction. The evaporation of the RG-O PIL suspension using a lens on a flat substrate with restricted geometry gives concentric rings (spherical lens) or stripe patterns (cylindrical lens) formed by the RG-O platelets that accumulate at the confined droplet edge during evaporation (**Figure 3.10**). The self-assembly process is driven by the combined effect of the controlled evaporation in a confined environment and the preferential interaction between RG-O PIL platelets and the charged substrate surface.

When the interaction with the substrate is not critical the method works very well as a general self-assembly process and well organized concentric rings of different types of nano-objects, such as nanoparticles,

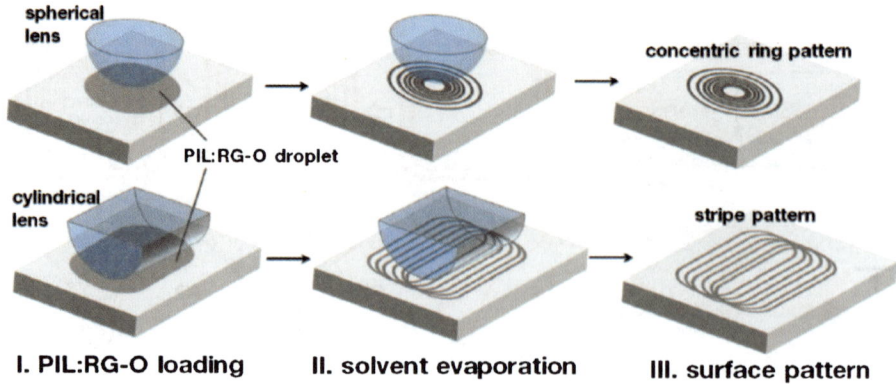

Figure 3.10   Scheme of formation of a self-assembled RG-O pattern by controlled
solvent evaporation. A homogeneous suspension of RG-O sheets
modified with a poly(ionic) liquid (PIL), poly(1-vinyl-3-ethylimidazo-
lium)bromide is placed in the microscopic gap between the curved upper
surface of a silica lens and the flat substrate. With evaporation of the
solvent the RG-O–PIL sheets deposit forming concentric (spherical
lens) or stripe (cylindrical lens) patterns. (Reprinted with permission
from T. Y. Kim *et al.*, *Adv. Mater.*, 2011, **23**, 2734. Copyright 2011,
John Wiley & Sons.)

carbon nanotubes[24] and nanowires[25] are easily self-ordered into
concentric gradient structures. The case of nanowires, in particular,
shows that some kind of controlled alignment of the nano-objects within
the rings can be achieved also *via* evaporation in a confined geometry.
The evaporation of a droplet of *N*,*N*′-dimethylquinacridone (DMQA), a
fluorescent dye that crystallizes into nanowires when the solution
concentrates during drying, leaves rings formed by packed and aligned
DMQA nanowires. The direction and rate of solvent evaporation is
controlled by a steady upstream of nitrogen gas in the evaporation room
(**Figure 3.11(a)**). After the solution dries on the substrate, ordered rings
of nanowires, which align and form concentric rings are observed
(**Figure 3.11(b)** and **(c)**). The distance between adjacent rings and the
density of every nanowire ring decrease from the external region toward
the center of the contact area giving, as in the general case of polymer
rings, a gradient of concentric patterns (**Figure 3.11(d)**). The concentration of DMQA affects the formation of ordered and aligned nanowires,
and a very dilute DMQA solution does not form concentric rings but
only a film with nanowires aligned in the direction of the solution flow
during evaporation.

The deposition method can be also used as a replica method, such as
the case of gold concentric rings previously cited, to self-organize
carbon nanotubes.[24] The process proceeds in two steps, in the first step

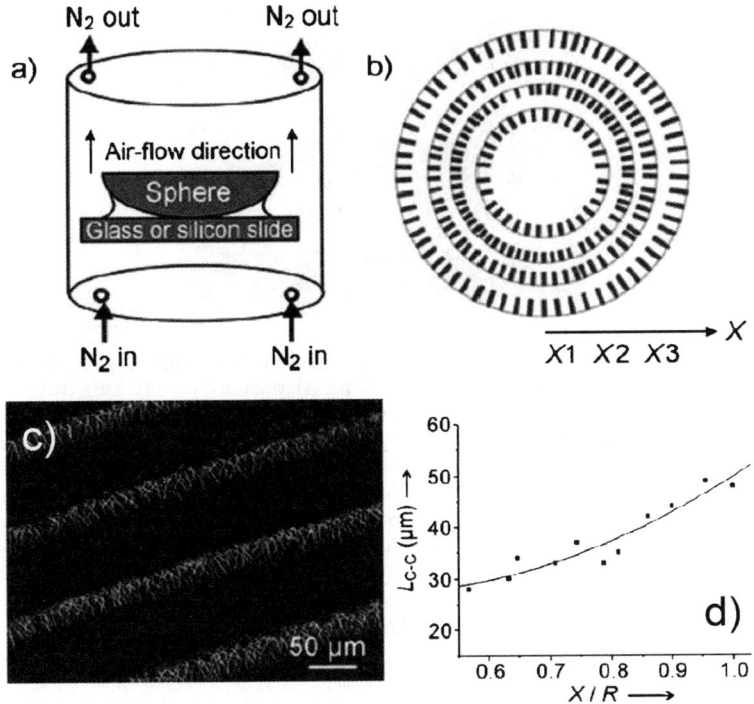

Figure 3.11 (a) Schematic set-up used for obtaining concentric rings of aligned nanostructures. (b) Concentric rings composed of organic nanowires. (c) SEM image of the DMQA nanowires in a ring, which form during solvent evaporation. (d) Spacing of the nanowire array as a function of the distance to the lens−substrate contact point. L_{C-C} is the distance between adjacent nanowires, X is the distance from the center of the lens−substrate contact and R is the radius of the outermost ring. (Reprinted with permission from Z. Wang *et al.*, *Angew. Chem., Int. Ed.*, 2011, **50**, 2811. Copyright 2011, John Wiley & Sons.)

gradient concentric rings on a silicon substrate are obtained using a linear conjugated polymer, poly[2-methoxy-5-(2-ethylhexyloxy)-1,4-phenylenevinylene] (MEH-PPV) through the sphere-on-substrate geometry. In the second step a droplet from an aqueous solution of multi-walled carbon nanotubes (MWNTs) containing a positively charged polyelectrolyte, poly(diallyl dimethylammonium) chloride, is cast onto the MEH-PPV ring-patterned silicon substrate. The MWNTs accumulate into the interspaces of the polymeric rings and when water evaporation has gone to completion they form a perfect replica. The different wettabilities of the periodically alternating hydrophobic MEH-PPV patterns and hydrophilic silicon interspaces facilitate the deposition. In the last processing step the MEH-PPV rings are removed

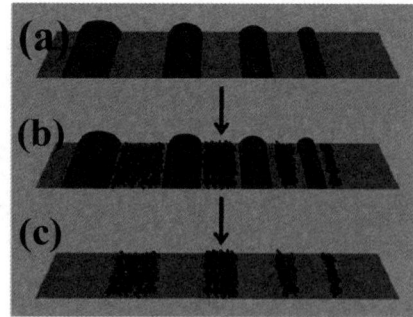

Figure 3.12 Fabrication of gradient concentric MWNT rings. (a) In the first step concentric MEH-PPV rings are deposited on Si substrate using a sphere-on-substrate geometry. The rings show a decrease in the distance between adjacent nanowires, L_{C-C}, and height, h, from outermost ring (left) toward the original sphere−Si contact center (right). (b) A drop of MWNT/PDDA water solution is cast onto the MEH−PPV ring-patterned Si substrate. Upon completion of water evaporation, MWNT rings form between the MEH−PPV rings. (c) Gradient concentric MWNT rings are finally obtained upon selective removal of MEH−PPV with toluene. (Reprinted with permission from S. W. Hong *et al.*, *Adv. Funct. Mater.*, 2008, **18**, 2114. Copyright 2008, John Wiley & Sons.)

by toluene leaving only the gradient concentric MWNTs rings (**Figure 3.12**).

REFERENCES

1. R. D. Deegan, *Phys. Rev. E*, 2000, **61**, 475.
2. M.E. R. Shanahan, *Langmuir*, 1995, **11**, 1041.
3. D. Orejon, K. Sefiane and M. E. R. Shanahan, *Langmuir*, 2011, **27**, 12834.
4. M. E. R. Shanahan, K. Sefiane and J. R. Moffat, *Langmuir*, 2011, **27**, 4572.
5. E. Rio, A. Daerr, F. Lequeux and L. Limat, *Langmuir*, 2006, **22**, 3186.
6. J. R. Moffat, K. Sefiane and M. E. R. Shanahan, *J. Phys. Chem. B*, 2009, **113**, 8860.
7. E. Adachi, A.S. Dimitrov and K. Nagayama, *Langmuir*, 1995, **11**, 1057.
8. R. D. Deegan, *Phys. Rev. E*, 2000, **61**, 475.
9. S. D. Iliev, *J. Colloid Interface Sci.*, 1999, **213**, 1.
10. L. Shmuylovich, A. Q. Shen and H. A. Stone, *Langmuir*, 2002, **18**, 3441.
11. S. Maheshwari, L. Zhang, Y. Zhu and H. C. Chang, *Phys. Rev. Lett.*, 2008, **100**, 4503.
12. L. Zhang, S. Maheshwari, H. C. Chang and Y. Zhu, *Langmuir*, 2008, **24**, 3911.
13. S. W. Hong, J. Xu and Z. Lin, *Nano Lett.*, 2006, **6**, 2949.
14. J. Xu, J. Xia and Z. Lin, *Angew. Chem.*, 2007, **119**, 1892.

15. S. W. Hong, M. Byun and Z. Lin, *Angew. Chem.*, 2009, **121**, 520.
16. Z. Lin and S. Granick, *J. Am. Chem. Soc.*, 2005, **127**, 2816.
17. S. W. Hong, J. Xu, J. Xia, Z. Lin, F. Qiu and Y. Yang, *Chem. Mater.*, 2005, **17**, 6223.
18. M. Byun, N. B. Bowden and Z. Lin, *Nano Lett.*, 2010, **10**, 3111.
19. J. Xu, J. Xia, S. W. Hong, Z. Lin, F. Qiu and Y. Yang, *Phys. Rev. Lett.*, 2006, **96**, 066104.
20. S. W. Hong, M. Byun and Z. Q. Lin, *Angew. Chem., Int. Ed.*, 2009, **48**, 512.
21. S. W. Hong, J. Xu and Z. Q. Lin, *Nano Letters*, 2006, **6**, 2949.
22. T. Y. Kim, S. W. Kwon, S. J. Park, D. H. Yoon, K. S. Suh and W. S. Yang, *Adv. Mater.* 2011, **23**, 2734.
23. Y. Zhu, S. Murali, W. Cai, X. Li, J. W. Suk, J. R. Potts and R. S. Ruoff, *Adv. Mater.*, 2010, **22**, 3906.
24. S. W. Hong, W. Jeong, H. Ko, M. R. Kessler, V. V. Tsukruk and Z. Lin, *Adv. Funct. Mater.*, 2008, **18**, 2114.
25. Z. Wang, R. Bao, X. Zhang, X. Ou, C.-S. Lee, J. C. Chang and X. Zhang, *Angew. Chem., Int. Ed.*, 2011, **50**, 2811.

CHAPTER 4

Convective Self-Assembly (CSA)

We have just seen how an ordered structure can be obtained through controlling the evaporation processes which appear basically as stochastic, such as the stick and slip motion. This order is restricted, however, to concentric structures and even if a wide variety of them can be easily obtained, linear parallel stripes remain difficult to fabricate. If we think carefully about the conditions, which are necessary to obtain such concentric structures, we observe that restricting the evaporation environment appears to be the main feature of the process. However, the rise of capillary forces at the boundary of the evaporating system plays another important role. What happens if the evaporating environment is constrained in a dynamic way, for instance creating a proper system for spreading the solution during evaporation? This is the subject of this chapter.

In the previous chapters we have observed how the evaporation of a colloidal droplet can be used for forming regular arrays mostly with a circular shape. The coffee stain and the Marangoni effect are the main physical−chemical driving forces that are behind this process. However, restrictions in the evaporation environment of a droplet can allow for the formation of concentric ordered structures, which are another step towards an advanced control of the process. What about organizing single or even multilayers of densely packed colloidal particles from the evaporating processes? This possibility is really interesting for several technological applications such as photonic crystals, antireflective coatings, surface-enhanced Raman spectroscopy (SERS) and several techniques have been developed so far. Colloidal films can be fabricated under controlled conditions by conventional film processing techniques, such as spin-coating and dip-coating, or by sedimentation, Langmuir−Blodgett, and microfluidic packing, just to

Water Droplets to Nanotechnology: A Journey Through Self-Assembly
By Plinio Innocenzi, Luca Malfatti and Paolo Falcaro
© P. Innocenzi, L. Malfatti and P. Falcaro 2013
Published by the Royal Society of Chemistry, www.rsc.org

cite some of the many possibilities. In the case of very thin films composed of one or a few more layers, evaporation in controlled conditions is an interesting option. We have shown in Chapter 3 that the formation of ordered rings by an evaporating droplet is dictated by the restriction of the evaporating surface with the rise of capillary forces at the liquid—air interface. This process is carried out under static conditions, meaning that the substrate and the material, which is used as the contact medium of the liquid, does not move during the entire evaporation process. A new coating technology is therefore necessary for depositing thin films of colloidal particles whilst being able to control the evaporation environment and interface at the contact line. This technique has to be specifically designed to exploit self-assembly driven by capillary forces.[1]

4.1 SELF-ASSEMBLY BY CONVECTIVE FLOW

In a free evaporating droplet, because of the coffee stain or Marangoni effect, internal flows of different types arise. If the evaporating environment is, however, physically restricted, the flows produce convective motions that are able to address self-organization onto a 3D solid substrate. Self-assembly by convective flow is particularly suitable if the particles are in the sub-micrometer range. A pre-requisite is the sedimentation or aggregation of particles remains negligible during the self-organization process. Another important point is the wettability of the substrate and indeed the contact angle of the droplet, which should not be larger than 30°. To achieve self-organization out of an evaporating colloidal droplet a confinement of the three-phase contact line has to be realized.[2,3] A typical experimental set-up[4] to achieve self-assembly by controlled convection at the triple line interface is as follows: above a horizontal solid substrate, which can move at a controlled speed, is an inclined plate at an acute angle while a small droplet of the colloidal solution is poured at the plate-substrate corner (**Figure 4.1**). The motion of the substrate at a controlled speed produces the dragging of the suspension, which leaves a thin evaporating liquid film behind. If the overall process is finely controlled, well-packed layers of the colloidal particles (the number of layers depends on the deposition parameters) self-organize on the substrate. This deposition technique is simple and the process well reproducible but what is the mechanism which forces dispersed colloidal particles to move and organize at the triple phase contact line of the droplet edge? If we have a closer look at the colloidal droplet between the two plates we can observe that it forms a concave meniscus and as a

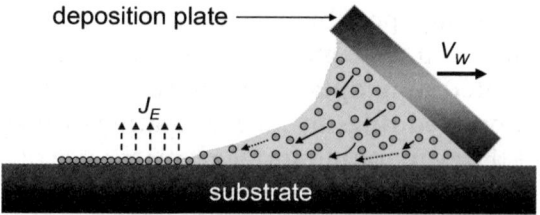

Figure 4.1 The scheme of deposition of colloidal films *via* CSA. The particles are driven by convective flow to the edge of the contact line where self-organize into an ordered layer. J_E is the evaporative flux of the liquid leaving the drying region and v_w the meniscus withdrawal speed. (Adapted with permission from B. G. Prevo *et al.*, *Langmuir*, 2004, **20**, 2099. Copyright 2004, American Chemical Society.)

consequence an attractive capillary force will arise. To compensate the solvent evaporation at the liquid-substrate contact line a flux of the liquid will be driven toward this drying region, such as in the case of the coffee stain. The convective flow induced by solvent evaporation drags the particles toward the capillary−solid edge. The assembly process, which at the end leaves an ordered array, starts when the thickness of the evaporating solvent film reaches the particle size[2].

The hydrodynamics of the overall process is quite complex but a simplified equation has been derived which describes the self-assembly of the particles as a function of flux in the drying region well in the steady state.[5] In the steady state the solvent evaporation is exactly compensated by the flow from the bulk suspension.

In the drying region of the thin evaporating liquid film the colloidal particles of diameter, h, deposit with a growth rate, v_c while the evaporating process in the same region is governed by three volumetric fluxes, J_E, which is the solvent evaporation flux, J_P, the particle flux and finally J_S the solvent flux to the drying (**Figure 4.2**). The meniscus moves at a rate v_w, which does

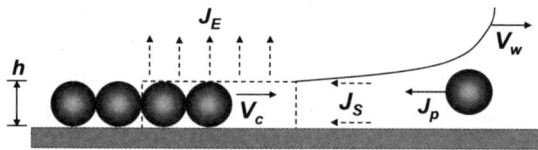

Figure 4.2 Schematic detail of the drying region during deposition of a colloidal film by CSA. The colloidal particles of diameter, h, deposit with a growth rate, v_c while the evaporating process in the same region is governed by three volumetric fluxes, J_E, the solvent evaporation flux, J_P, the particle flux and finally J_S the solvent flux to the drying region; the meniscus moves at a rate v_w. (Reprinted from B. G. Prevo *et al.*, *Curr. Opin. Colloid Interface Sci.*, **311**. Copyright 2007, with permission from Elsevier.)

not necessarily coincide with the particle growth rate, v_c. In general measuring J_E is quite a complicated process because evaporation is not constant along the entire drying region and is a function of the distance from the meniscus. The simplest way to describe the assembly is, therefore, to incorporate all the convective contributions of the molecular transport into a unique hydrodynamic parameter, K. The simplified formula,[6] which describes the deposition kinetics of the films becomes:

$$v_C = \frac{K\Phi}{h(1-\varepsilon)(1-\Phi)} \qquad (4.1)$$

where ε is the porosity of the deposited colloidal film and Φ the volume fraction of the particles in suspension. On the basis of eqn (4.1), which works well under steady-state conditions, when will a uniform crystalline monolayer form? If the evaporation rate and volume fraction remain fixed the equation predicts that there is only one very specific speed at which a crystalline monolayer of defined thickness forms. This happens only when the influx of particles toward the drying region exactly compensates for the particles consumption due to the layer growth, which means that the crystal growth rate should be equal to the substrate speed (v_s), $v_c = v_s$.

It is interesting to observe that this method allows for depositing uniform layers of particles of different sizes, from microspheres to nanoparticles and moreover the particles form crystalline structures. Within some limits, controlling the type of order and the number of layers by the deposition speed is also possible. In general, the thickness of the colloidal film can be only an integer multiple of the diameter of a single particle. The number of layers depends on the substrate velocity, the concentration of the particles in the colloidal solution and the meniscus shape[7]. If the substrate speed is increased above the assembly rate for a monolayer, only an incomplete layer forms. This happens also if, equivalently, the volume fraction of the suspension is reduced. On the other hand, a decrease of the substrate speed or an increase of the volume fraction gives rise to multiple layers.

An example of the phases obtained by combining different deposition speeds and particle volume fractions is given in the 'phase diagram' drawn by Prevo and Velev,[4] which shows the number and type of layers which are obtained as a function of suspension volume fraction, Φ, and deposition speed, v_w (**Figure 4.3**). The lower right region (high volume fraction and low deposition speed) corresponds to the conditions that give multilayer colloidal films while the top left region (low volume fraction and high deposition speed) represents the combination which gives incomplete monolayers (submonolayers). Now we have an

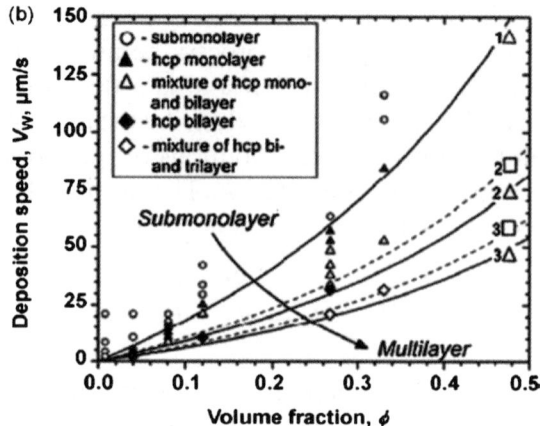

Figure 4.3   The deposition of 1.1 μm monodispersed polystyrene particles at 30% relative humidity. Changing the volume fraction and deposition speed allows obtaining an 'operational phase diagram'. The number of layers and the packing structures are plotted as a function of suspension volume fraction, Φ, and deposition speed, v_w. The solid curves indicate the natural assembly speeds, v_c, for layers of different type calculated from eqn (4.1). For all curves only one fitting parameter K is used. The predicted v_c for the square or assembly speeds are shown with dashed lines to underline their metastable, transition behavior. (Reprinted from B. G. Prevo *et al.*, *Curr. Opin. Colloid Interface Sci.*, **311**. Copyright 2007, with permission from Elsevier.)

interesting question to answer: how do the colloidal particles pile up in multilayers? The more stable packing arrangement is hexagonal as this is the most favorable phase from a thermodynamics point of view because of the small free volume which is left. Intermediate layers, however, such as the square phase, or cubic phases can form and appear as transition 'metastable' states to the stable hexagonal phase.[8]

In general, considering the evaporation rate as being constant during the process is a good approximation. This is a rough simplification but it works quite well to satisfy the terms in eqn (4.1). However, if the temperature of the substrate could be adjusted in a controlled way this represents a quite efficient way of controlling the evaporation rate.[9] Deposition *via* CSA of 500 nm polystyrene particles in water on a movable substrate with a heater has shown that the crystal growth rate, v_c, increases with the concentration of the particles volume fraction Φ, as described in eqn (4.1), but also non-linearly with the substrate temperature, T_s, (**Figure 4.4**). This result shows that the convective flow and, therefore, the growth rate created by evaporation can be tuned by adjusting the substrate temperature.

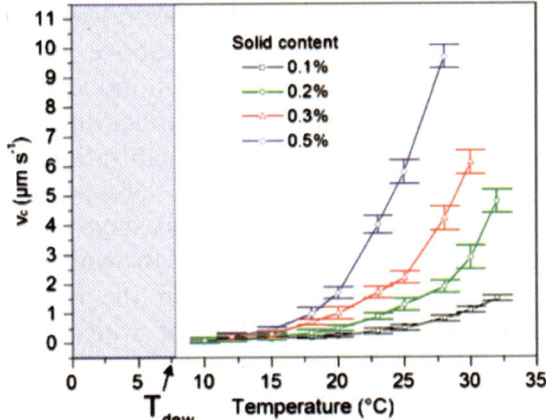

Figure 4.4 Dependence of the crystal growth rate, v_c, on the surface temperature T_s for different particle volume fractions, Φ, ranging from 0.1 to 0.5%. The experiment was performed using an aqueous solution of 500 nm diameter polystyrene particles. The ambient temperature and relative humidity (RH) were 20 °C and 42%, respectively, and T_{dew} was 7.7 °C. (Adapted with permission from L. Malaquin *et al. Langmuir*, 2007, **23**, 11513. Copyright 2007, American Chemical Society.)

What are the advantages of CSA with respect to a 'traditional' dip-coating process? The deposition speed is a clear advantage because using solutions with high-volume fractions allows for increasing the rate of film formation. In the dip-coating process quite low-volume fractions are generally employed and this is reflected by lower crystallization rates. If we look again at the data in **Figure 4.3** we can clearly see the correlation between the deposition rate and the increase of concentration in the colloidal solution. This, however, has a limit. The concentration of the colloidal particles cannot in fact increase over the value $\Phi \sim 0.58$, which is the volume fraction at which a transition to a glassy state is observed.[10] There is also another important point to note here and this is the degree of order that is achievable by this method. The colloidal films have a polycrystalline structure and a perfectly ordered monolayer, but because of the intrinsic nature of the drying process, which goes through the creation of multiple nucleation and growth sites, this is impossible to achieve.

4.2 REVERSING SELF-ASSEMBLY

The evaporation rate and the particle concentration, as just observed, govern the crystal growth rate, but changing the temperature during the deposition of a colloidal film *via* CSA can be surprising. For an aqueous solution the T_{dew} is a temperature limit for the system because below this threshold value (temperature and pressure) of the system water vapor condenses. If the temperature is above the T_{dew}, nothing unexpected is observed and the evaporation-induced convective flow creates a particle motion toward the contact line. Decreasing the temperature clearly slows down the self-organization and when $T_s \approx T_{dew}$ the evaporation in the drying region self-assembly almost stops because no particle flux is generated.[9] At this stage the particles do not move on average with the exception of Brownian motion. If the temperature goes below the dew point, in the former drying zone condensation starts and a reverse flow of water from the already assembled layer to the bulk liquid gives a progressive disassembly of the monolayer (**Figure 4.5**). This process is clearly reversible and the

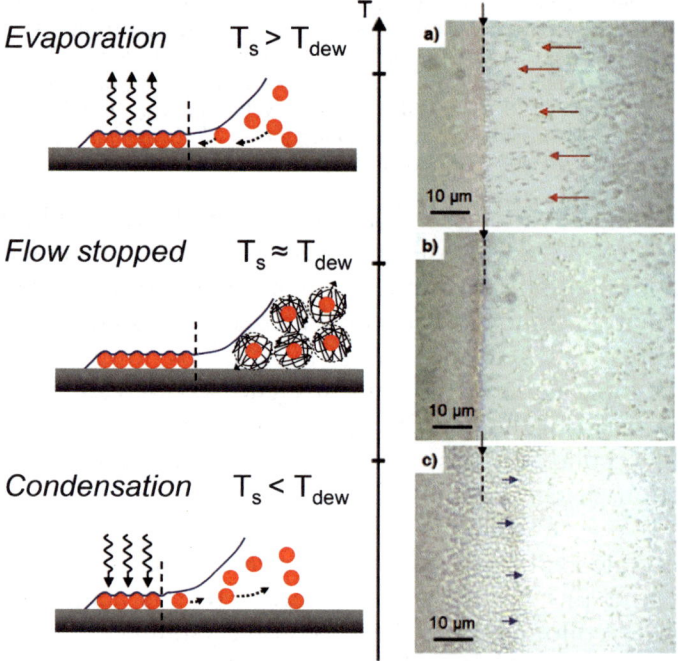

Figure 4.5 Evaporation-induced convective flows as a function of temperature with 500 nm polystyrene particles (0.2% solid content) on an oxygen-plasma-treated PDMS surface. (a) $T_s > T_{dew}$, self-assembly of a particle monolayer. The particles accumulate to the contact line drag by solvent flow. (b) $T_s \sim T_{dew}$, self-assembly stops. The evaporation of the solvent in the drying region is almost zero and because particle flux is created, the assembly process is stopped. Only Brownian motion of the particles is observed. (c) $T_s < T_{dew}$, reversing self-assembly. Condensation takes place on the already assembled layer and creates a reverse flow of solvent that disassembles the monolayer. (Adapted with permission from L. Malaquin *et al.*, *Langmuir*, 2007, **23**, 11513. Copyright 2007, American Chemical Society.)

colloidal layers can be disassembled and reassembled several times without affecting the structure of the lattices.

4.3 CONVECTIVE DEPOSITION OF BINARY SUSPENSIONS

The deposition method is not restricted to a single-component colloidal solution and it works well also in depositing binary mixtures in a single step. In this case the dimension of the two types of particles and the volume fraction ratio become very important. In general, a good strategy for an efficient packing of the monolayer is to use particles of two different ranges, nano and micro. This allows the particles in the nanoscale to occupy the small interstitial spaces left free by the microspheres, which pack in an ordered structure driven by the capillary forces during evaporation. If organic polymers make up the microparticles they can also be quite easily removed after deposition of the layer, which acts at any effect as a templating agent. An example is the binary solution formed by gold nanoparticles (10–25 nm) and latex microspheres (400–1000 nm).[11]

An ordered colloidal crystal array of closely packed latex spheres is formed close to the three-phase contact line by CSA while at the same time the smaller gold nanoparticles fill the void space within the colloidal crystal (**Figure 4.6**). This means that the well-packed colloidal crystal layer is also a template for the gold nanoparticles, which aggregate in the voids.[12] After removal of the templating latex microspheres by solvent or pyrolytic etching what remains is an inverse replica of the colloidal crystal template. These films have a good field of application as SERS substrates due to their highly porous and ordered structure.[13,14] CSA can be also used to deposit binary solutions containing silver nanoparticles, which increase the Raman signal and the species to be detected, such as bacteria.[15]

Another possible application that has been demonstrated for binary colloidal solutions deposited *via* CSA is the fabrication of antireflective coatings.[16] Coatings prepared from silica nanoparticles with bimodal size distribution ($\sim$70 and $\sim$130 nm) have shown a nearly 90% reduction in the single side reflectance for films deposited from suspensions of $\sim$1 : 1 large–small particles. The antireflective properties are enhanced with respect to layers of single components because the smaller particles disrupt the structure of the larger particles with a decrease in the packing fraction.

Another interesting example is the structure which is obtained by deposition of an aqueous binary suspension of 1 μm silica microspheres

One-step template-directed assembly

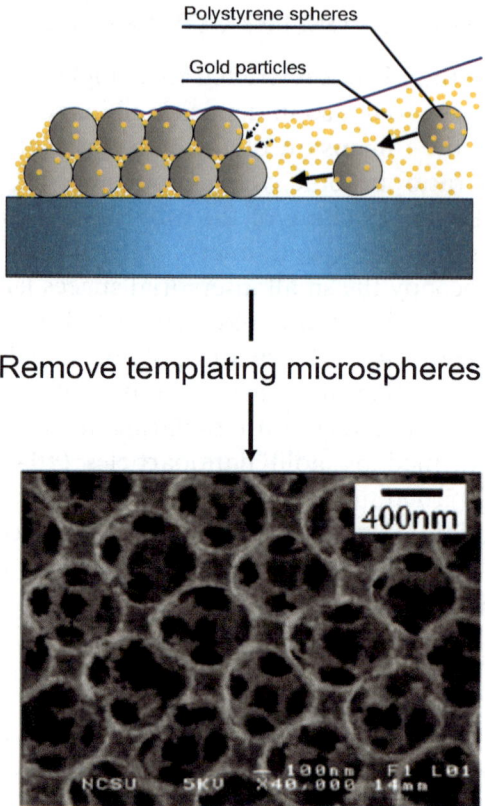

Figure 4.6 One-step template directed fabrication of SERS substrates by CSA
deposition of a binary mixture of gold nanoparticles and polystyrene
microspheres (top drawing). After removal of the polystyrene template a
gold replica of the microparticle array is obtained. (Reproduced with
permission from ref. 12.)

and 100 nm polystyrene (PS) nanoparticles.[17] After deposition by CSA
a monolayer is formed with the larger silica microspheres that are
partially submerged in an array of smaller polystyrene (PS) nanoparticles. Neighboring microspheres are in contact and nanoparticles fill
the interstitial regions up to a height that is lower than the microsphere
diameter and depends on the relative concentration of the nano and
micro particles in solution (**Figure 4.7**).

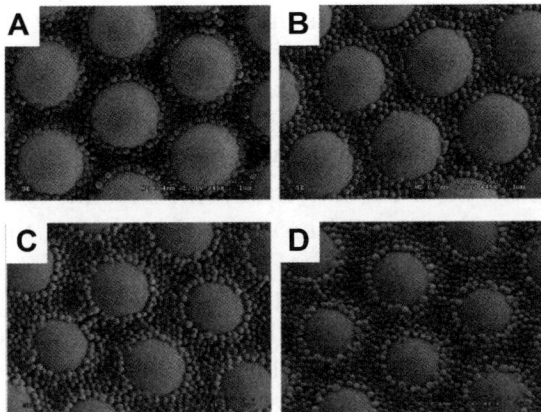

Figure 4.7 SEM images of the morphologies that result after deposition by CSA of a binary solution of microspheres and nanoparticles; the microspheres appear partially buried by nanoparticles. The samples have been obtained from solutions with Φ micro = 0.2 and Φ nano= 0.02 (A), 0.04 (B), 0.10 (C) and 0.12 (D). Close microspheres are in contact with each other while the nanoparticles fill the interstitial space without full coverage of the microspheres. (Adapted with permission from P. Kumnorkaew *et al.*, *Langmuir*, 2009, **25**, 6070. Copyright 2009, American Chemical Society.)

4.4 STICK AND SLIP AGAIN! GETTING ALIGNED STRIPES OF NANOPARTICLES FROM CSA

CSA clearly appears as a competitive deposition method for the fabrication of ordered colloidal films but the real challenge is getting not only ordered planar layers but specific structures, such as aligned stripes, without employing pre-patterned substrates. To understand how this is possible *via* CSA let us return to the stick and slip effect. We are now familiar with stick and slip and we know that if controlled during the evaporation of a droplet it can become a tool for achieving ordered structures. What happens during CSA is not that different from coffee stain with the exception of the rise of a capillary force at the edge of meniscus, which drives the particle organization during drying. Stick and slip should, therefore, be expected and linear stripes parallel to the triple contact line should be observed at least in some conditions. The mechanism of stripe formation is the same as that in evaporating colloidal droplets, the controlled linear movement of the substrate, however, this allows for achieving regularly spaced particle stripes. Stick and slip, in any case, is not always observed during CSA because it is dependent on the evaporation rate and substrate speed. The stick and slip stage starts when the particles are transported by the convective flow towards the meniscus where they organize. In the slip stage the movement of the substrate causes the elongation of the meniscus, which finally detaches from the monolayer. The repetition of this process gives the formation of aligned wires, which form an ordered array.[18] There is, however, a limit in this order because the array periodicity cannot be really controlled. This happens because a direct correlation between the speeds at which the meniscus moves and the sticking time that exists. A decrease of the meniscus speed increases the sticking time, which also means the width and thickness of the particle stripes increase. In a thicker stripe the pinning force of the meniscus is larger and therefore the stick stage also becomes longer because the liquid film can elongate more. After depinning the contact line moves to a longer distance and so the next stripe distance will also be longer.

A solution to get regularly spaced stripes, which is based on a 'stop-and-go' CSA, allows for a precise control of the distance between the parallel wires.[19] Instead of moving the substrate with a constant speed, the substrate is moved quickly and then stopped at a fixed position (**Figure 4.8**) for a given time. The meniscus is then again translated with high speed to the next 'stop' position at a distance D, leaving a particle wire of certain width, w, and thickness h. This method works very well and parallel wires with the same width and distance can be obtained (**Figure 4.9**). High sensitivity strain gauges of close-packed 14 nm gold nanoparticles on flexible polyethylene terephthalate films have been fabricated using this approach.[20]

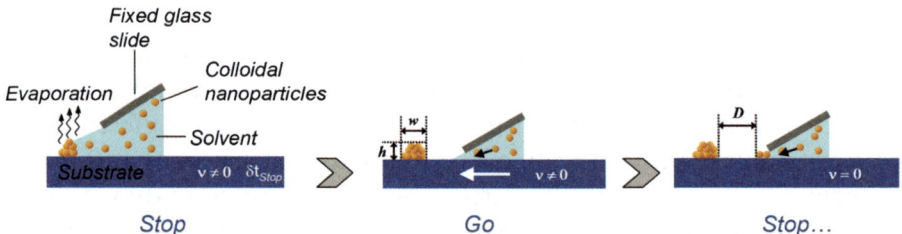

Figure 4.8   Deposition of a gold nanoparticle linear array by a modified CSA in a stop-and-go mode. The deposition is realized through a sequence of stop and go steps: (i) in the first step the meniscus is quickly moved to a fixed position and then kept for a given time interval; (ii) in the second 'go' step the meniscus is again translated with high speed v, leaving a deposited nanoparticle wire of width w and thickness h; (iii) the meniscus in the second 'stop' stage is again arrested after a chosen distance D. This process can be repeated several times to give sequence of nanoparticle-aligned wires. (Adapted with permission from C. Farcau *et al.*, *ACS Nano*, 2011, **5**, 7137. Copyright 2011, American Chemical Society.)

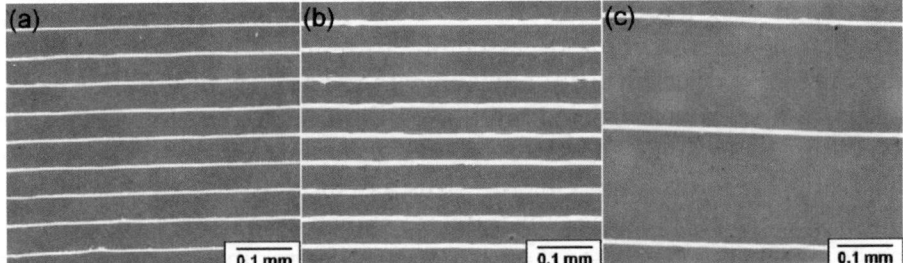

Figure 4.9 Optical microscopy images of arrays of identical parallel wires made of 14 nm gold nanoparticles using stop-and-go CSA. (a) The wires have a width, w, of 6 μm and are separated by a distance, D, of 0.05 mm; (b) w = 9 μm and D = 0.05 mm and (c) w = 9 μm and D = 0.02 mm. The method allows for the deposition of wires, which have the same height and the same distance between each other. (Adapted with permission from C. Farcau *et al.*, *ACS Nano*, 2011, **5**, 7137. Copyright 2011, American Chemical Society.)

The possibility of forming ordered stripe arrays by CSA is not only limited to directions parallel to the triple contact line; perpendicular wires form under the same conditions when the temperature of the substrate is lowered as this produces a reduction of solvent convective flow. This phenomenon can be quite surprising because it does not appear as a fingering instability phenomenon, in fact the wires do not form all together, they form at different times. The low flow reduces the movement of the particles toward the meniscus, which do not pin on the substrate but preferentially concentrate at the meniscus edge. Pinning is triggered only when, locally, a critical particle density is achieved and the tip of the wire moves out perpendicularly to the meniscus. This event is triggered several times and gives the formation of regular NPs stripes. Furthermore, by changing the substrate temperature, the orientation of the wires can be switched from parallel to perpendicular to the substrate−liquid−air contact.

4.5 NOT ONLY PARTICLES!

Spherical nano and microparticles are typical isotropic objects that form ordered layers during deposition by CSA. This method also works surprisingly well in the case of anisotropic colloids of a different nature. Biological colloids, in particular, offer a large number of examples of objects with a wide range of aspect ratios. Viruses, for instance, have an aspect ratio of 17 for the tobacco mosaic virus (TMV) and 147 for the bacteriophage M13, and furthermore, they are strictly monodispersed. Deposition by CSA of rod-shaped TMV viruses produce films with a high degree of long-range order.[21] During deposition, TMVs align parallel to the direction of assembly forming monolayers of some centimeters in length. By adjusting the TMV concentration and assembly speed, well-ordered multilayers from 2 to 12 can be obtained. The top virus layer, however, shows different degrees of in-plane disorder. The morphology of the virus assembly is also dependent on the surface energy of the substrate, as shown in **Figure 4.10**. Hydrophilic surfaces generate more ordered layers in comparison to hydrophobic ones. Highly ordered domains can be also obtained by CSA from icosohedral virus-like particles (VLPs) that form oriented 2D superlattices. Analysis *in situ* of the self-assembly process by grazing-incidence small-angle X-ray scattering has shown that in the case of an aqueous solution, a transport-limited assembly, where convective flow directs assembly of VLPs into a lattice orientation with respect to the water drying line, is at the origin of the process.[22]

Bacteria provide another example of biological colloidal solutions used in CSA. *E. coli*, a Gram-negative bacilli, and *Staphylococcus cohnii*, a Gram-positive coccus, have been co-deposited with silver

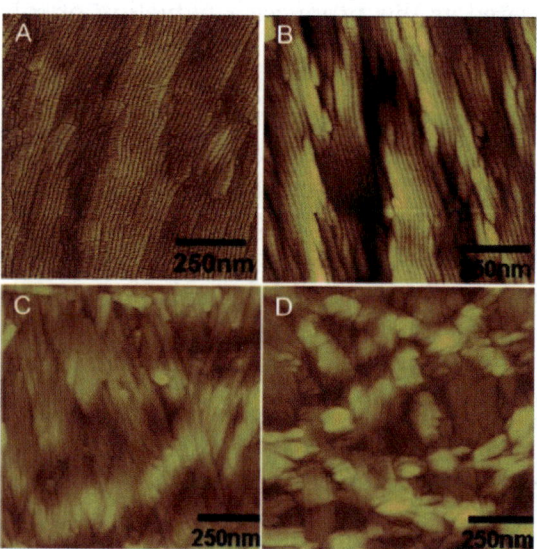

Figure 4.10 AFM images of TMV films assembled onto a silicon substrate which as been modified by amines (A), native silica on silicon (B), acryloxy modified (C), and methyl modified (D). The TMV have been deposited using a solution with a 0.009 volume fraction at a speed of 10 μm s⁻¹ and 10% RH (*Z* scale = 75 nm). The average surface roughness is 4.26, 7.54, 10.68 and 17.14 nm for images (A)–(D), respectively. (Reprinted with permission from S. P. Wargacki *et al.*, *Langmuir*, 2008, **24**, 5439. Copyright 2008, American Chemical Society.)

nanoparticles, as already described in the case of binary systems, and used for SERS analysis.[15]

Besides pre-formed objects, such as particles and biological colloids, another interesting option is to use CSA for the deposition of films whose structure will be formed during evaporation and drying. This strategy has been applied for producing chitosan films[23] and ultra-thin mesoporous silica layers through evaporation-induced self-assembly.[24] The evaporation drives the formation and organization of micelles that act as templates of mesopores. This subject is the topic of Chapter 8.

Different interesting examples of other applications of CSA also exist.[25–29]

REFERENCES

1. P. A. Kralchevsky and N. D. Denkov, *Curr. Opin. Colloid Interface Sci.*, 2001, **6**, 383.
2. E. Adachi, A. S. Dimitrov and K. Nagayama, *Langmuir*, 1995, **11**, 1057.
3. B. G. Prevo, D. M. Kuncicky and O. D. Velev, *Colloids and Surfaces A: Physicochem. Eng. Aspects*, 2007, **311**, 2.
4. B. G. Prevo and O. D. Velev, *Langmuir*, 2004, **20**, 2099.
5. A. S. Dimitrov and K. Nagayama, *Chem. Phys. Lett.*, 1995, **243**, 462.
6. The first formula has been proposed in: A. S. Dimitrov and K. Nagayama, *Langmuir*, 1996, **12**, 1303.
7. P. Born, S. Blum, A. Munoz and T. Kraus, *Langmuir*, 2011, **27**, 8621.
8. N. Denkov, O. Velev, P. Kralchevski, I. Ivanov, H. Yoshimura and K. Nagayama, *Langmuir*, 1992, **8**, 3183.
9. L. Malaquin, T. Kraus, H. Schmid, E. Delamarche and H. Wolf, *Langmuir*, 2007, **23**, 11513.
10. D. Quemada and C. Berli, *Adv. Colloid Interface Sci.*, 2002, **98**, 51.
11. P. M. Tessier, O. D. Velev, A. T. Kalambur, J. F. Rabolt, A. M. Lenhoff and E. W. Kaler, *J. Am. Chem. Soc.*, 2000, **122**, 9554.
12. D. M. Kuncicky, B. G. Prevo and O. D. Velev, *J. Mater. Chem.*, 2006, **16**, 1207.
13. P. M. Tessier, O. D. Velev, A. T. Kalambur, A. M. Lenhoff, J. F. Rabolt and E. W. Kaler, *Adv. Mater.*, 2001, **13**, 396.
14. D. M. Kuncicky, S. D. Christensen and O. D. Velev., *Appl. Spectrosc.*, 2005, **59**, 401.
15. M. Kahramam, M. M. Yazici, F. Sahin and M. Culha, *Langmuir*, 2008, **24**, 894.
16. B. G. Prevo, Y. Hwang and O. D. Velev, *Chem. Mater.*, 2005, **17**, 3642.
17. P. Kumnorkaew and J. F. Gilchrist, *Langmuir*, 2009, **25**, 6070.
18. C. Farcau, H. Moreira, B. Viallet, J. Grisolia and L. Ressier, *ACS Nano*, 2010, **4**, 7275.

19. C. Farcau, N. M. Sangeetha, H. Moreira, B.Viallet, J. Grisolia, D. Ciuculescu-Pradines and L. Ressier, *ACS Nano*, 2011, **5**, 7137.
20. C. Farcau, H. Moreira, B. Viallet, J. Grisolia, D. Ciuculescu-Pradines, C. Amiens and L. Ressier, *J. Phys. Chem. C*, 2011, **115**, 14494.
21. S. P. Wargacki, B. Pate and R. A. Vaia, *Langmuir*, 2008, **24**, 5439.
22. C. E. Ashley, D. R. Dunphy, Z. Jiang, E. C. Carnes, Z. Yuan, D. N. Petsev, P. B. Atanassov, O. D. Velev, M. Sprung, J. Wang, D. S. Peabody and C. J. Brinker, *Small*, 2011, **7**, 1043.
23. S. F. Maloy, G. L. Martin, P. Atanassov and M. J. Cooney, *Langmuir*, 2012, **28**, 2589.
24. Z. Yuan, D. B. Burckel, P. Atanassov and H. Fan, *J. Mater. Chem.*, 2006, **16**, 4637.
25. Z. Yuan, D. N. Petsev, B. G. Prevo, O. D. Velev and P. Atanassov, *Langmuir*, 2007, **23**, 5498.
26. A. D. Ormonde, E. C. M. Hicks, J. Castillo and R. P. Van Duyne, *Langmuir*, 2004, **20**, 6927.
27. L. Ressier, B. Viallet, A. Beduer, D. Fabre, L. Fabie, E. Palleau and E. Dague, *Langmuir*, 2008, **24**, 13254.
28. B. Viallet, L. Ressier, L. Czornomaz and N. Decorde, *Langmuir*, 2010, **26**, 4631.
29. K. Chen, S. V. Stoianov, J. Bangerter and H. D. Robinson, *J. Colloid Interface Sci.*, 2010, **344**, 315.

CHAPTER 5
Using Breath for Nanotechnology

In the previous four chapters we have discovered that an evaporating
liquid droplet is a highly technological laboratory where sophisticated
self-assembly phenomena can occur. However, this is not the only nano-
facility available in every day life. Every time you breathe on a cold and
flat surface, for example, you produce a sort of micro-pattern that is
formed by small water droplets that could be exploited for the
fabrication of films and membranes that are structured at the
nanoscale. This chapter deals with this phenomenon, the so-called
'breath figure', from the simplest examples to the latest applications in
nanotechnology.

5.1 THE NANO-SHAPING BREATH

Children often breathe on glass windows and write words on them using
their fingers (**Figure 5.1**). The letters disappear immediately after the
glass becomes transparent and this makes the kids very happy with their
own way of exchanging 'secret messages'. Despite the poor efficacy of
this 'encryption' method, breathing on a flat surface is a simple but
effective way of producing lithographic-free patterns from the micro- to
the nano-scale. One of the first scientists who recognized that the
manner in which vapor condenses upon ordinarily clean surfaces of
glass or metal was not only a childish amusement, was Lord John
William Strutt Rayleigh who, in a letter to the journal *Nature*, observed
that 'the condensed water is in the form of small lenses, often in pretty
close juxtaposition'.[1] Because of its mode of generation, Lord Rayleigh
called the arrangement of water droplets 'breath figures'. The
condensation of dew on a surface is not *per se* an interesting tool to
design and organize matter but, if we manage to induce the same

Water Droplets to Nanotechnology: A Journey Through Self-Assembly
By Plinio Innocenzi, Luca Malfatti and Paolo Falcaro
© P. Innocenzi, L. Malfatti and P. Falcaro 2013
Published by the Royal Society of Chemistry, www.rsc.org

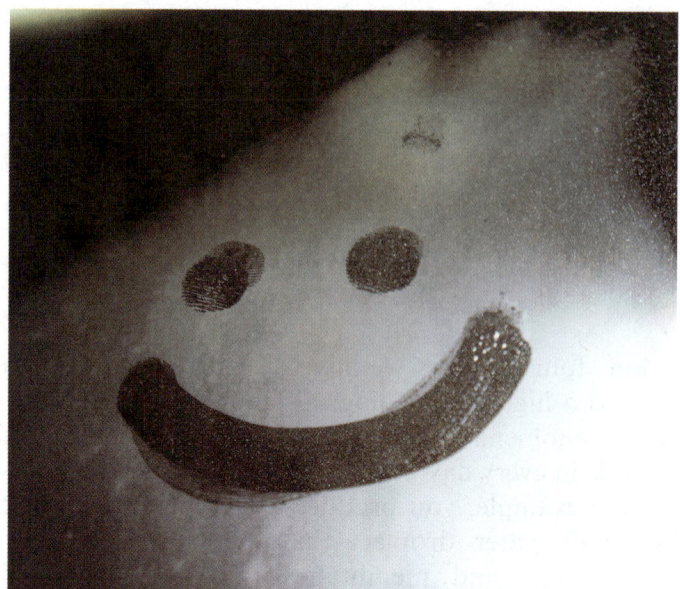

Figure 5.1 Breath figures on a cold window. (The photograph was taken by L. Malfatti. P. Falcaro, A. Pinna and B. Lasio contributed to the artwork with their creativity and breath.)

phenomenon in liquids, the process starts increasing its appeal. This is exactly what Knobler and Beysens did in 1988 as they studied the formation of breath figures on liquid paraffin.[2] Interestingly they discovered that, when condensed on oil, the water droplets did not immediately coalesce, but rather they self-organized into a 2D 'close-packed-like' hexagonal packing. They attributed this organization to the presence of a thin-oil layer, which did not allow two nearby droplets to coalesce immediately into a bigger droplet.

The next step towards the development of a templating method for materials was the formation of breath figures on liquid layers, which hereafter we define as *underlayers*, made by polymers dissolved in a volatile solvent over which the water droplet can condense. The strategy is simple and can be achieved in five stages: (1) a cold surface is created by solvent evaporation from the underlayer; (2) environmental moisture condenses at the underlayer–atmosphere interface; (3) the condensed water droplets move and self-organize into islands with hexagonal packing; (4) the condensed water droplets start sinking into the film acting as moulds for pores; and (5) total evaporation of the polymer solvent and water reveal the final porous film.

The first example of this approach dates back to 1994 when a flux of water vapor was applied to a mixture of star polystyrene and poly(*para*-phenylene)-*block*-polystyrene dissolved in carbon disulfide (CS_2). As a consequence of the CS_2 evaporation, the surface of the polymer—solvent film becomes cold and pushes the moisture to condense in the form of droplets. In this system, the liquid water, to some extent, also produces an imprint of the polymeric layer below giving rise, after solvent evaporation, to a honeycomb porous structure with a pore diameter of a few micrometers.[3] Since these pioneering works, the use of breath figures has been extended not only to polymers, but also to other types of materials, such as ceramics,[4] proteins[5] and sol—gel based films.[6] Many studies have also been devoted to the control of the porous structure, at least in a limited range, obtaining the remarkable goal of tuning the droplets, and therefore the pore size, from a few hundreds of nanometers up to 10 micrometers.[7,8]

5.2 THE TEMPLATING MACHINERY

Until now we have described the templating induced by breath figures as a simple and effective technique, but, as it often occurs in non-equilibrium phenomena, fine control of all the parameters, which affect the *machinery* of the process, is a hard task to achieve. First of all, nowadays, it is still not clear if a general mechanism, applicable to all the systems where breath figures occur, could exist. Secondly, the degree of order of the water template, final pore size distribution and pore organization are critically affected by the chemical composition of the underlayer over which the water droplets start condensing.[9] Despite the difficulty of implementing a general theory for the understanding of the phenomenon, it is possible, however, to draw some general conclusions from the large variety of studies published so far.

The formation of pores on the final materials is dependent on the presence of the condensed water droplets *in* or *onto* the underlayer during solvent evaporation. If the water remains on the surface, only a superficial 2D porous structure is obtained. However, if the water droplets enter into the evaporating layer, the final material shows a 3D porosity. How is this possible? There are several pieces of evidence that point out how the water droplets *sink* into the underlayer leaving the surface available for inducing further water droplet formation by breath figures.[10] Once the droplets are immersed in the evaporating layer, however, they remain isolated and do not merge inducing phase separation. In addition, in most of the cases, the buried water droplets, or rather water 'bubbles', are monodispersed and tend to form an ordered array of remarkable extent. Once the 3D-organized array is formed, some connections between bubbles can be established, leading to an interconnected porous structure.

Three main issues arise from the experimental evidence we have presented: (a) What is the driving force of sinking? (b) Why do the bubbles not merge? (c) Why do the droplets, at least in some cases, self-organize into long-range ordered structures?

All the answers to these questions deal with a few main phenomena and, despite the complexity, we will make an attempt to explain them briefly.

(a) Water sinking can be caused by two main phenomena: Marangoni convective motion and thermocapillarity of the evaporating underlayer. The Marangoni effect was extensively explained in Chapter **2**. Solvent evaporation induces a temperature variation in some parts of the underlayer, with the upper layers being colder than the lower layers. This causes a difference in layer density and, therefore, a circular motion from the top to the bottom of the film as shown in **Figure 5.2**. The mass flux drags the water droplets that condensed on the surface towards the inner part of the underlayer causing sinking. The second phenomenon, which is maybe less known, is due to the difference in the surface tension of two liquids as a function of temperature. Some authors have proposed the z_0 parameter for predicting the formation of 2D or 3D porous structures by evaluating the interfacial energy between the solvent of the underlayer and the water.[11] z_0 can be calculated using this formula:

$$z_0 = (\gamma_W - \gamma_{W/U})/\gamma_U \tag{5.1}$$

where γ_W and γ_U are the surface tension of the water and the

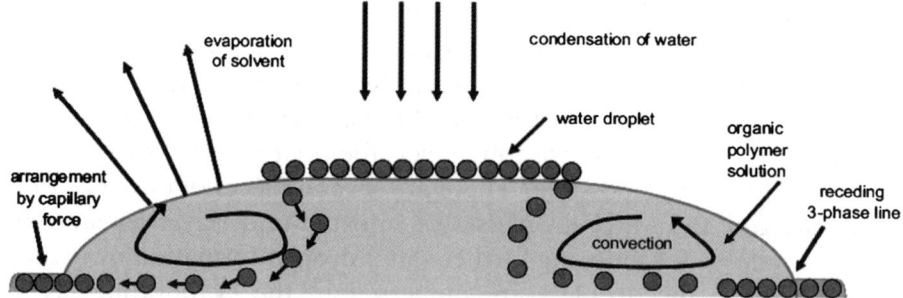

Figure 5.2   Mechanism of breath figure formation. (Reprinted from N. Maruyama *et al.*, *Thin Solid Films*, **327**, 854. Copyright 1998, with permission from Elsevier.)

underlayer, respectively, and $\gamma_{W/U}$ is the interfacial tension between the water and the underlayer. If the value of z_0 is between -1 and 1 the condensed droplets will float between the air and the underlayer. However, if $z_0 > 1$ the droplets will swell below the surface producing a 3D porosity. Despite the importance of this model, the formula requires the surface tension values values of the polymer mixture, which have to be measured and, in particular, we should note that either γ_U and $\gamma_{W/U}$ change as a function of the solvent evaporation in the underlayer. This means that, at least in some cases, the sinking driving force is time-dependent and the value can continuously change until the underlayer hardens and motion freezes.[12]

(b) As we have already noticed, the most fascinating side of breath-figure formation is that the condensed water droplets remain separated and do not merge completely. The explanation of this behavior is still under debate but, however, at least two different explanations have been provided so far. The first one assigns the stability of the droplet-ordered domains on the underlayer to the combined effect of thermocapillarity and Marangoni convection. These two contributions allow the droplets to not be in contact with each other at least until the last stages of the templating process. Following this explanation, the water droplets levitate on a buffer layer made by solvent-saturated air over the underlayer.[10] Under these conditions, Marangoni convection also plays an important role because it allows the droplets to be kept separate if a sufficient thermal gradient is obtained.

Other researchers have provided a second explanation that implies the formation of a membrane that encapsulates the droplets once they have formed on the underlayer surface.[13] The encapsulation of the water droplets into a membrane made by a drying layer of soft matter (polymer of gel) would allow for insulating the bubbles, preventing them from merging, while the sinking should be attributed to Marangoni motion and thermocapillarity.

(c) The arrangement of water bubbles into a long-range ordered array is undoubtedly a complex balance between competing repulsive and attractive forces, leading to a surface filled with stabilized water droplets.[14] The mechanism is still not clear, but some observations are worth looking at. By following *in situ* the formation and organization of the breath figures with laser scattering, some researchers have observed that the arrangement of the droplets was irregular until the coverage of the underlayer was nearly complete (**Figure 5.3**).[15,16] At this stage Marangoni convection and the thermocapillarity effect play an important role in determining the order. However, these two effects are not the only ones responsible for organization and some other

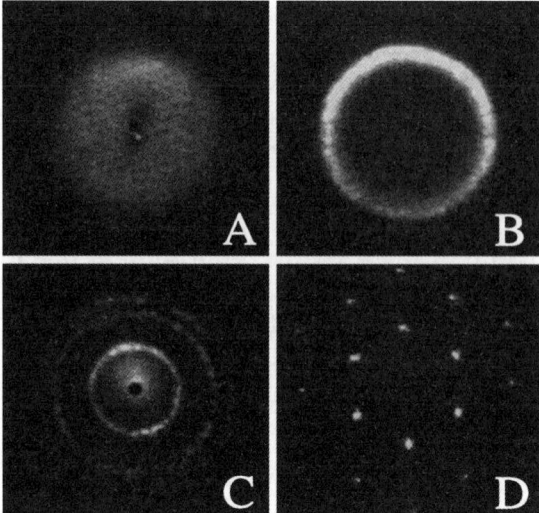

Figure 5.3   *In situ* observation, formation, and organization of the breath figures with
laser scattering. (Reprinted with permission from ref. 16. Copyright 1999,
with kind permission from Springer Science and Business Media.)

important parameters have to be considered. The sinking process, for
instance, seems to be critical for obtaining well-organized porous
structures. In fact, if the droplet sinking occurs too early during the
drying process, the bubbles do not have time to reach a monodispersed
dimension, producing a broad pore-size distribution.[14] There are several
parameters that affect droplet sinking and these will be described in
more detail in Section 5.3.

The substrate over which the evaporating underlayer is deposited also
gives a contribution to the long-range order formation, even though
until now it has been the object of few studies. For instance, it has been
observed that a higher degree of pore order is obtained if a polymer
solution is cast on a freshly cleaved mica surface with respect to the pore
organization obtained on a glass substrate.[17] The role of the substrate is
correlated to the wettability of the evaporating underlayer with respect
to the surface over which it is deposited: the higher the wettability, the
higher the extent of the ordered porous structure.[18]

5.3   WHICH PARAMETERS CONTROL THE BREATH FIGURES
TEMPLATING PROCESS?

Like most of the processes described in this book, the breath figure is
extremely easy to observe in everyday life and triggering the
phenomenon requires minimal equipment, however, if a full control

of the process is necessary, a large variety of free parameters have to be controlled. Pore dimension, pore-size distribution, degree-of-order and, in some cases, pore shape, are largely dependent on these parameters. By changing the environmental conditions such as relative humidity, airflow or temperature, considerably different results can be obtained. The most effective parameters, which control breath-figure formation, are listed below.

Humidity

The most obvious free parameter is the moisture, *i.e.* the amount of water vapor used to induce condensation of the water droplets. Water vapor can be introduced both under static and dynamic conditions. In the first case it is important to stress that breath figures do not form in an atmosphere that has less than 45–50% of relative humidity.[9] Higher relative humidity values induce more regular pore arrays but also cause an almost linear increase of the water droplet size.[13,19]

Alternatively, the relative humidity over the underlayer can be dynamically changed by a moist airflow. These configurations allow tuning other control parameters such as humidity content, flow rate, distance, and angle of direction of the airflow.[20] The flux of air onto the underlayer induces a fast evaporation of the solvent and strongly lowers the temperature of the surface creating a bigger temperature gradient with respect to the bulk. This, in turn, causes faster condensation of the water droplets and stronger convective motions, which lead to close-packed water-droplet domains on the underlayer.

Temperature

As a consequence of solvent evaporation, it has been observed that the surface temperature of the underlayer quickly drops from ambient temperature to $-4\ °C$.[21] This is the first driving force for condensation of the water droplets. Therefore, controlling the temperature of the evaporating underlayer (and the substrate) allows tuning of the breath-figure formation, as well as enhancement of the condensation and the degree of order. Some research groups designed a specific cold-casting cell to control the temperature below ambient conditions. In general, lower temperatures cause an increase in the viscosity of the underlayer and change its interfacial energy with the condensing water droplets. By controlling the environmental temperature between 3–5 °C, for example, poly(p-phenylenevinylene) and polythiophene-made membranes exhibiting a regular porosity extended for tens of mm^2 were prepared.[22]

Vacuum

This control parameter acts mainly on the solvent evaporation rate. This causes a faster fall of the temperature and strongly boosts the breath-figure formation. Unlike temperature, however, the environmental pressure does not increase the viscosity of the underlayer and therefore the convective motion necessary for self-assembly of the water droplet is not influenced.

Solvent

One of the most critical choices for the design of a breath-figure process is the solvent selection. We have already mentioned CS_2, which works very well and is considered the 'default' solvent, at least for polymers. The solvent cannot be chosen, however, without considering the type of polymers, or other materials, that it will be mixed with. Chloroform, dichloromethane, freon-type solvents, benzene, xylene, tetrahydrofuran and mixtures of solvents have been used for breath-figure formation depending on the chemical composition of the underlayer to be templated. In general, a slow evaporation rate of the solvents allows for the sinking of the water bubbles and therefore should promote the formation of a 3D pore array in the materials.[23]

Underlayer Concentration

The concentration of the underlayer can be used as a tool to control, within a certain range, the pore size of the final materials. It has been found that higher concentrations are responsible for smaller pore dimensions and this can be generally explained considering the increased capability of a concentrated underlayer to stabilize the surface energy of tiny water droplets. For polymers, the following equation has been proposed, which describes the pore size (PS) variation as a function of the concentration (c):

$$PS = k/c \tag{5.2}$$

where k is a constant that is dependent on the material used.[24] Some studies also have shown that a higher concentration of the underlayer may induce the formation of a 3D porous structure while less concentrated systems would lead to 2D pore organization.[25]

Chemical Composition of the Porous Materials

The breath figure templating approach can be theoretically applied to all types of materials that can be dissolved in organic and volatile solvents. The most obvious, and studied, system is that of polymers and many publications have focused on the changes in the pore organization as a function of the chemical composition, molecular weight and chemical modification.[9,14] Although a defined rule regulating the effect of the polymer architecture and microstructure on the porous structure does not exist, the quality of the pore array strictly depends on the mixture of materials and solvents. Polymers can also be used to form suspensions of nano-objects made by organic,[26] inorganic[27,28] or hybrid organic−inorganic[29] materials. The application of the breath figure templating approach to this mixture allows for fabricating functional and organized coatings and membranes.

5.4 SHAPING PORE SIZE AND PORE ORGANIZATION

One of the remarkable features of the breath figure templating approach is the possibility of tuning the pore size over two orders of magnitude: relative humidity, and concentration of the underlayer are the most used parameters for tuning the pore dimension. Even though the smallest theoretical size of a water droplet at room temperature has been calculated as 10 nm,[30] to date very few papers have reported the synthesis of a film with a pore size smaller than 100 nm.[31] However, the preparation of bigger pores is limited by the coalescence of micron-sized water droplets.[14] Besides this, breath figure allows the production of very large ordered arrays of droplets that are monodispersed over a large area, generally 4–50 mm^2, but in some cases up to several cm^2.

The droplet lattices often present some defects, which are due to isolated vacancies of bubbles (something similar to Frenkel or Schottky defects in ionic crystals), or to boundaries among adjacent droplet domains with the same symmetry but different geometrical orientation.[9] The polydispersity of the droplets also causes defects in the organized arrays forming larger micron-scaled macropores surrounded by a corona of smaller voids.[32]

Until now, we have generally mentioned pore organization as a 2D or 3D close-packed array. In most cases the water droplets self-organize towards the formation of so-called *honeycomb* structures where the symmetry of the pore planes parallel to the substrate is hexagonal, as depicted in **Figure 5.4**; however other types of pore arrangement can be obtained using particular techniques. If the templating breath-figure

Figure 5.4   Honeycomb structure obtained by the breath figures templating approach. (Reprinted by permission from Macmillan Publishers Ltd: G. Widawski *et al., Nature*, 1994, **369**, 387, copyright 1994.)

approach is applied to a viscoelastic polymer, for instance, it is sufficient to stretch or compress the porous film producing a pore array with rectangular, squared or triangular symmetry, as shown in **Figure 5.5**.[33]

Besides mechanical deformation, a different symmetry of the water droplet arrays is obtained using an inclined substrate. This experimental set-up leads, at first, to a distorted square or hexagonal symmetry depending on the stacking of adjacent rows.[34]

5.5   BREATH FIGURES AT WORK: APPLICATIONS OF TEMPLATED POROUS MATERIALS

Because of their structure, porous films or membranes obtained by breath figures may find potential applications in filtration,[35] antireflective and superhydrophobic surfaces,[36,37] cell culturing and scaffolds for tissue engineering,[38] fuel cells,[39] bioassays,[40] and soft-lithography.[11] In this section we will focus on some examples that involve organization at the nanometer scale.

The addition of nano-objects into the polymer–solvent mixture allows at first for the preparation of a homogeneous dispersion and then the formation of porous films with functional nano-structures that are well dispersed inside the materials. In some cases, the organized porosity of the coating provides special optical properties to the materials, which behave as photonic crystals (see Chapter 6). When the optical properties, due to the porous morphology, are merged with

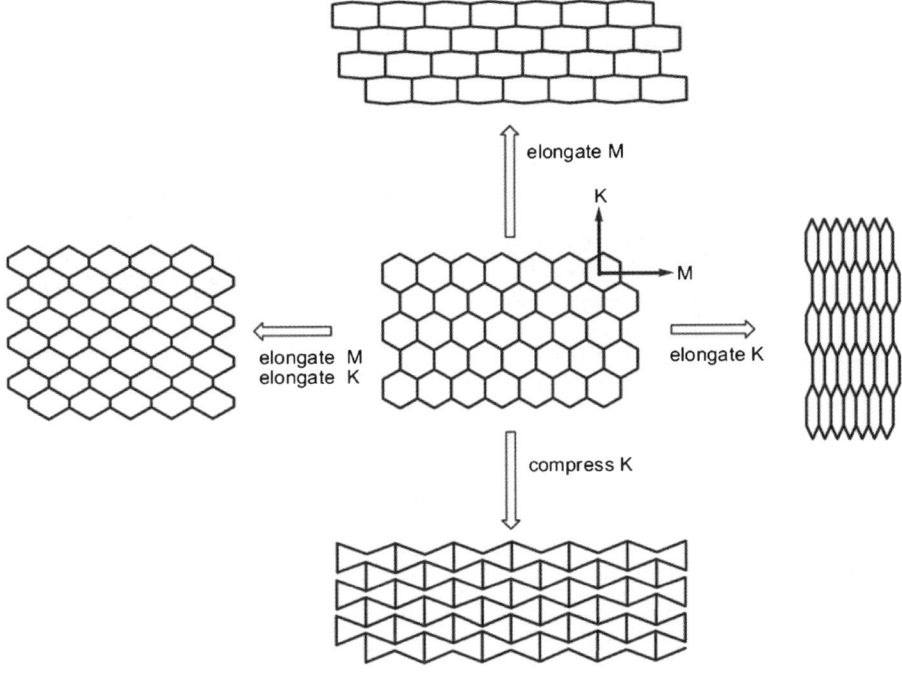

Figure 5.5 Effects of mechanical deformation of a honeycomb pattern as a function of the elongation direction. (Reprinted with permission from U. H. F. Bunz, *Adv. Mater.*, 2006, **18**, 973. Copyright 2006, John Wiley & Sons.)

those of the functional nanostructures, the material becomes a promising candidate for organic electronic devices with applications in microelectronic or photovoltaic devices. This is the case for a breath figure-templated coating made by conjugated polymers, mixed with organic conductive nanoparticles such as CNTs[41] (**Figure 5.6**) or C_{60} fullerenes.[42] In both cases, the addition of the nanostructures resulted in a quenching of the fluorescent properties of the bare film due to charge transfer. This is reliable evidence that both nano-objects are not aggregated and can efficiently interact with the polymeric matrixes. The same direct and easy approach has been also extended to the homogeneous distribution of gold and silver nanoparticles, which conferred plasmonic properties to the porous materials.[26]

A more intriguing application of breath figures involves the formation of the so-called 'Pickering emulsion', which is an emulsion stabilized by the presence of solid particles.[43] During the breath-figure formation, the condensation and sinking of water droplets push the nanoparticles to stabilize the hydrophilic phase by decorating the water bubbles. After evaporation of the water, a sharp segregation of the

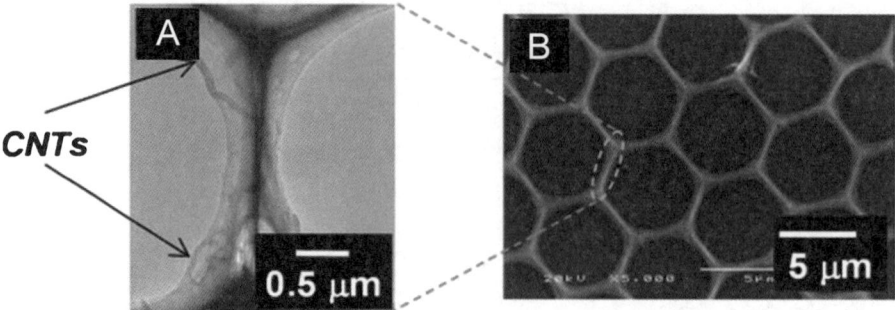

Figure 5.6   TEM and SEM images showing the morphology of organic hybrid film
containing carbon nanotubes. (Reprinted with permission from M. H.
Nurmawati *et al.*, *Adv. Funct. Mater.*, 2006, **16**, 2340. Copyright 2006,
John Wiley & Sons.)

nanostructures has been observed on the pore surface. This approach
has been used to order several types of nanoparticles such as zeolite
crystals,[44] quantum dots (CdSe, and core shell CdSe/CdS),[45,28]
magnetic and superparamagnetic (Fe_3O_4 and Fe_2O_3)[46] or plasmonic
nanoparticles (Au, Ag).[26] In most cases the segregation of the
nanostructures at the pore surface has been observed by confocal
microscopy or by TEM as shown in **Figure 5.7(a)** and **(b)**, respectively.
A research group from Zhejiang University has been able to control the
localization of silica nanoparticles within a honeycomb film by
changing the surface functionalization of the nanostructures. The
addition of hydrophilic or hydrophobic SiO_2 nanoparticles did not
influence the breath figure templating process but, while the first objects
self-organized at the pore surface, the other objects remained in the pore
walls (**Figure 5.8**).[47]

Some researchers have used the 2D hexagonal porous layer to induce
in situ nucleation and growth of nanostructures. The precursors of the
nanoparticles, have both been reduced during film drying, to form
noble metal clusters,[48] or hydrolyzed and polycondensed after film
pyrolysis (ZnO and TiO_2).[6,49] Thermal treatment does not dismantle the
porous architecture and leaves a final ceramic structure.

The pyrolitic treatment of polymeric coatings containing zinc oxide
and CNTs precursors have been also been used as seeding substrates for
the growth of nanorods and nanotubes by hydrothermal treatment or
chemical vapor deposition, respectively.[50] The nanostructures grown on
the substrate maintained their honeycomb structure, as shown in
Figure 5.9.

Sanchez and co-workers have reported the synthesis of porous
inorganic layers using an inorganic suspension of 'nano building blocks'

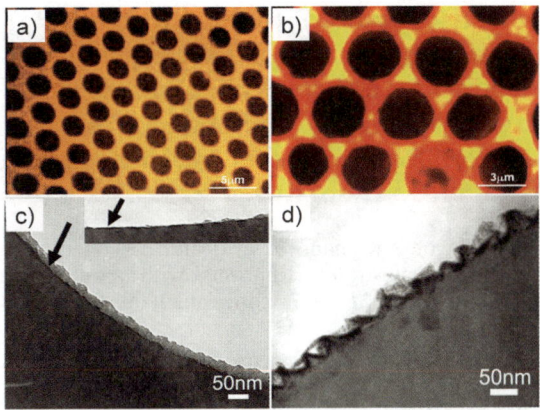

Figure 5.7 (a) and (b) Confocal microscopy of phase separated zeolite crystal on the pore surface. (c) and (d) TEM images of a porous polystyrene film containing CdSe nanoparticles at the pore surface. (Figures (a) and (b) reprinted with permission from V. Vohra *et al.*, *Langmuir*, 2009, **25**, 12019. Copyright 2009, American Chemical Society., Figures (c) and (d) reprinted by permission from: A. Boker *et al., Nat. Mater.* 2004, **3**, 302. Copyright 2004, Macmillan Publishers Ltd.)

(NBBs) made of silica, titania, Co and CdS nanoparticles.[51] Remarkably, the suspensions did not contain any polymer scaffolds of the dried coating as different types of surfactant were used instead to stabilize the water droplets formed by breath figures. Despite a lack of organization of the pores, a hierarchical structure containing two orders of porosity was obtained. The assembly of NBBs into the inorganic network allowed for the formation of smaller mesopores with a 10 nm size, while larger pores of tens of micrometers were produced by breath figures.

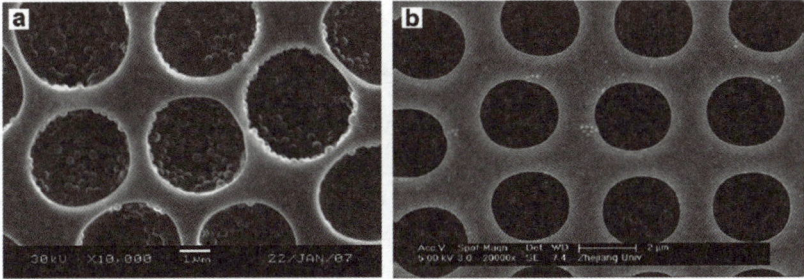

Figure 5.8 SEM images of film fabricated from solutions prepared by adding hydrophilic or hydrophobic silica particle in alcoholic suspension. (Reprinted from W. Sun *et al.*, *Polymer*, **51**, 4169. Copyright 2010, with permission from Elsevier.)

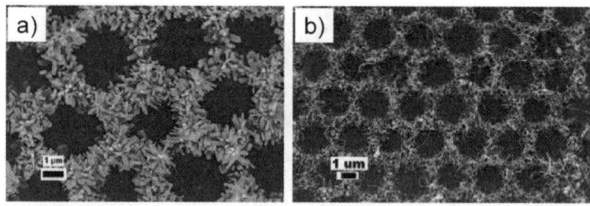

Figure 5.9   SEM images of ZnO nanorods (a) and carbon nanotubes (b) grown of porous seeding substrates. (Reprinted with permission from L. Lin *et al.*, *Chem. Mater.*, 2009, **21**, 4977. Copyright 2009, American Chemical Society.)

It is well known that functionalization and modification, on a microscale, of flat coatings gives rise to special surface properties, such as hydrophobicity.[52] Indeed, if the wettability of the surface is very low, a cast water droplet does not spread on it but, on the contrary, it rolls away such as a rigid sphere. This phenomenon, generally known as the 'lotus effect' can be also obtained by breath figures templating because 2D hexagonal arrays of micropores give to the surface a suitable roughness and, in addition, many types of polymers are hydrophobic. The superhydrophobic response can be further enhanced by peeling the pores arrays with adhesive tape. This treatment allows for producing a pin cushion structure with submicrometer pillars, as shown in **Figure 5.10**. By using this approach, Shymomura and co-workers have been able to obtain a coating with a water contact angle close to 170°.[53] By changing the chemical composition of the polymer structure, the pin cushion film can also show a reversible pH responsive property that makes the surface hydrophilic or hydrophobic after immersion into pH 3 or pH 9 buffer solutions for 1 h.[54] This surface morphology is very attractive for different types of applications and has been also used as a template for the synthesis of zinc oxide or silver 'nano-spikes' using electro-less or vacuum deposition, respectively.[55,56]

5.6   BREATH FIGURES AND SUPERSTRUCTURES

Nowadays the need for implementing applications based on nanomaterials is pushing the research community to combine a *top-down* approach, such as lithography or moulding, with *bottom-up* techniques, such as breath figures or other self-assembly processes. This explains the efforts made by researchers to create *superstructures*, which are materials of different lengths, based on breath figure-templated coatings. The first explorative experiments have been focused on the possibility of depositing porous films on *non-flat* surfaces such as a TEM grid. A polymeric mixture was at first cast on grids with different

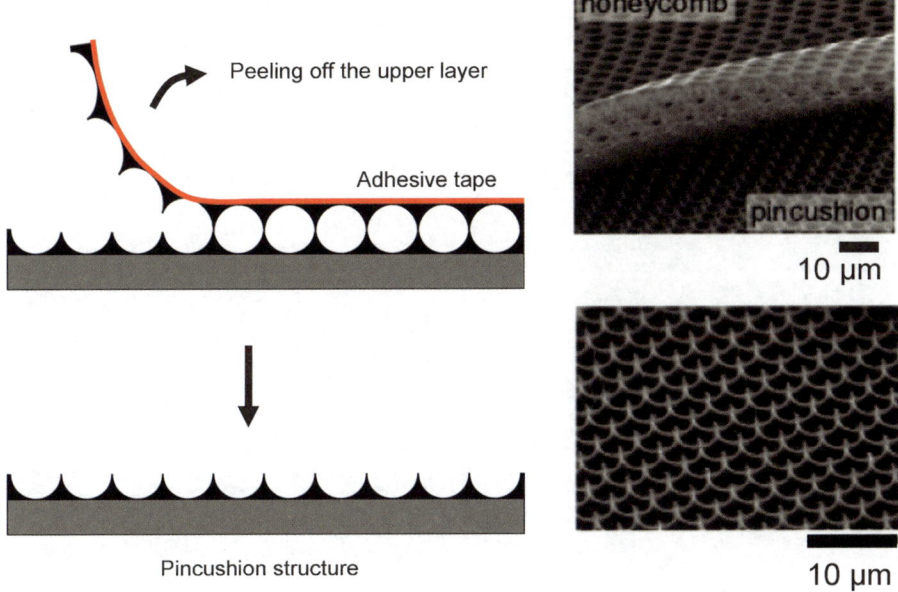

Figure 5.10   Schematic illustration and SEM images of the process for fabricating superhydrophobic pincushion structures. (Reprinted with permission from H. Yabu *et al.*, *Langmuir*, 2005, **21**, 3235. Copyright 2005, American Chemical Society.)

mesh sizes and then templated by breath figure.[57] It was found that only particular polymers that had a soft and flowing nature enabled the replication of the complex contours of the TEM grids forming highly ordered honeycomb coatings. Afterwards, similar superstructures were also used as a template for the preparation of moulded stamps with polydimethylsiloxane (PDMS), a kind of silicone rubber.[58] Superstructures of porous films have also been obtained through the UV exposure of photocurable materials. In this technique, breath-figure templating is applied to a photoresponsive molecule leading to the formation of a 2D hexagonal porous structure, and then the resulting structure is patterned by UV lithography through a photomask to induce selective chemical crosslinking of the exposed regions. After developing arrays tens of micrometers in width, the resulting patterns were found to show a porous honeycomb structure with a pore size of around 700 nm (**Figure 5.11**).[59]

More recently, a slightly different approach involving the addition of photoresponsive organic molecules (spiropyrans) to a mixture of polymer and solvent has been used.[60] After formation, the films were exposed to UV lithography causing the conversion of hydrophobic

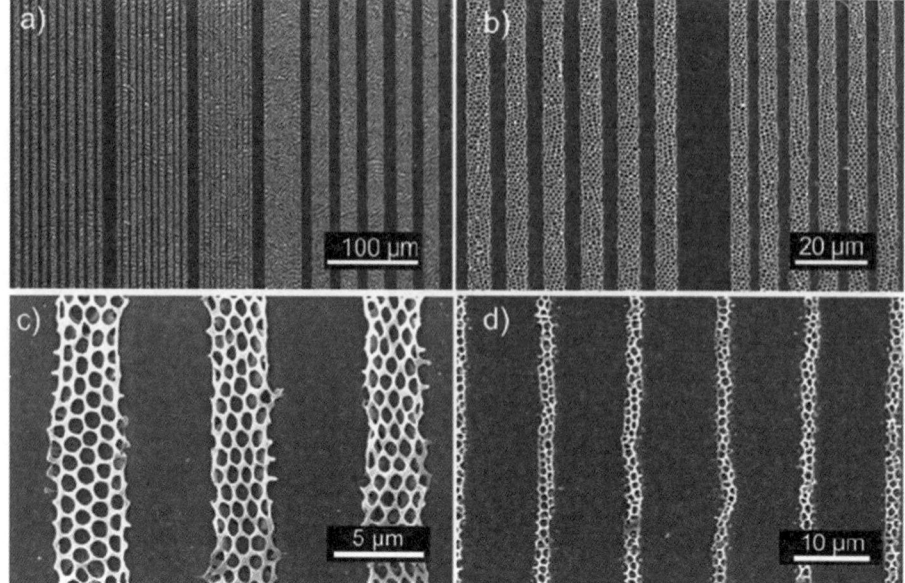

Figure 5.11 FE-SEM images (a)−(d) of the patterned porous superstructures obtained by UV photolithography. (Reprinted with permission from J. H. Kim *et al.*, *Adv. Mater.*, 2009, **21**, 4130. Copyright 2009, John Wiley & Sons.)

spiropyrans into hydrophilic merocyanines. By exploiting the solubility difference between these two isomers, patterning of the porous film was obtained. The non-irradiated parts were selectively dissolved by exposing the whole film to chloroform vapors.

Full control of the porous structure and geometry can also be obtained by forcing breath-figure formation in some underlayer regions. Kim and co-workers obtained this interesting result using the physical confinement of various shaped gratings put over the evaporating mixture of the polymer. This technique led to microarrays of porous film with a dimension of tens of micrometers and a long-range ordered organization of the breath figure-induced porosity of a few micrometers.

REFERENCES

1. J. W. S. Rayleigh, *Nature*, 1911, **86**, 416.
2. C. M. Knobler and D. Beysens, *Europhys. Lett.*, 1988, **6**, 707.
3. G. Widawski, M. Rawiso and B. François, *Nature*, 1994, **369**, 387.
4. B. C. Englert, S. Scholz, P. J. Leech, M. Srinivasarao and U. H. F. Bunz, *Chem. − Eur. J.*, 2005, **11**, 995.

5. Y. Zhang and C. Wang, *Adv Mater.*, 2007, **19**, 913.
6. K. Kon, C. N. Brauer, K. Hidaka, H.-G. Löhmannsröben and O. Karthaus, *Langmuir*, 2010, **26**, 12173.
7. O. Pitois and B. Francois, *Eur. Phys. J. B*, 1999, **8**, 225.
8. H. T. Lord, J. F. Quinn, S. D. Angus, M. R. Whittaker, M. H. Stenzel and T. P. Davis, *J. Mater. Chem.*, 2003, **13**, 2090.
9. U. H. F. Bunz, *Adv Mater.*, 2006, **18**, 973.
10. M. Srinivasarao, D. Collings, A. Philips and S. Patel, *Science*, 2001, **292**, 79.
11. A. Bolognesi, C. Mercogliano, S. Yunus, M. Civardi, D. Comoretto and A. Turturro, *Langmuir*, 2005, **21**, 3480.
12. P. Escalé, L. Rubatat, L. Billon and M. Sauve, *Eur. Polym. J.*, 2012, **48**, 1001.
13. N. Maruyama, T. Koito, J. Nishida, T. Sawadaishi, X. Cieren, K. Kjiro, O. Karthaus and M. Shimomura, *Thin Solid Films*, 1998, **327**, 854.
14. M. H. Stenzel, C. Barner-Kowollik and T. P. Davis, *J. Polym. Sci. Part A: Polym. Chem.*, 2006, **44**, 2363.
15. O. Pitois and B. Francois, *Eur. Phys. J. B*, 1999, **8**, 225.
16. O. Pitois and B. Francois, *Colloid Polym. Sci.*, 1999, **277**, 574.
17. L. Ghannam, M. Manguian, J. Francois and L. Billon, *Soft Matter*, 2007, **3**, 1492.
18. X. F. Li, Y. Wang, L. A. Zhang, S. X. Tan, X. L. Yu and N. Zhao, *J. Colloid Interface Sci.*, 2010, **350**, 253.
19. M. Hernandez-Guerrero, T. P. Davis, C. Barner-Kowollik and M. H. Stenzel, *Eur. Polym. J.*, 2005, **41**, 2264.
20. K. H. Wong, M. Hernandez-Guerrero, A. M. Granville, T. P. Davis, C. Barner-Kowollik and M. H. Stenzel, *J. Porous Mater.*, 2006, **13**, 213.
21. T. Nishikawa, J. Nishida, R. Ookura, S.-I. Nishimura, S. Wada, T. Karino and M. Shimomura, *Mater. Sci. Eng. C*, 1999, **8**, 495.
22. V. L. Govor, I. A. Bashmakov, R. Kiebooms, V. Dyakonov and J. Parisi, *Adv. Mater.*, 2001, **13**, 588.
23. L. Billon, M. Manguian, V. Pellerin, M. Joubert, O. Eterradossi and H. Garay, *Macromolecules*, 2009, **42**, 345.
24. M. H. Stenzel, *Aust. J. Chem.*, 2002, **55**, 239.
25. C. Cheng, Y. Tian, Y. Shi, R. Tang and F. Xi, *Macromol. Rapid Commun.*, 2005, **26**, 1266.
26. M. H. Nurmawati, P. K. Ajikumar, R. Renu and S.Valiyaveettil, *Adv. Funct. Mater.*, 2008, **18**, 3213.
27. W. Sun, Z. Shao and J. A. Ji, *Polymer*, 2010, **51**, 4169.
28. H. M. Ma, J. W. Cui, J. F. Chen and J. C. Hao, *Chem. − Eur. J.*, 2011, **17**, 655.
29. G. Vamvounis, D. Nystrom, P. Antoni, M. Lindgren, S. Holdcroft and A. Hult, *Langmuir*, 2006, **22**, 3959.
30. C. Kittel and H. Kroemer, *Thermal Physics*, Freeman, New York, 1980.
31. T. Hayakawa and S. Horiuchi, *Angew. Chem., Int. Ed.*, 2003, **42**, 2285.

32. M. S. Barrow, R. L. Jones, J. O. Park, M. Srinivasarao, P. R. Williams and C. J. Wright, *Spectroscopy (Amsterdam, Neth.)*, 2004, **18**, 577.
33. T. Nishikawa, N. Nonomura, K. Arai, J. Hayashi, T. Sawadaishi, Y. Nishiura, M. Hara and M. Shimomura, *Langmuir*, 2003, **19**, 6193.
34. O. Karthaus, X. Cieren, N. Maruyama and M. Shimomura, *Mater. Sci. Eng. C*, 1999, **10**, 103.
35. C. Greiser, S. Ebert and W. A. Goedel, *Langmuir*, 2008, **24**, 617.
36. M. S. Park and J. K. Kim, *Langmuir*, 2005, **21**, 11404.
37. Y. C. Chiu, C. C. Kuo, C. J. Lin and W. C. Chen, *Soft Matter*, 2011, **7**, 9350.
38. T. Nishikawa, J. Nishida, R. Ookura, S.-I. Nishimura, S. Wada, T. Karino and M. Shimomura, *Mater. Sci. Eng. C*, 1999, **10**, 141.
39. N. Zhang, J. Li, D. Ni and K. Sun, *Appl. Surf. Sci.*, 2011, **258**, 50.
40. M. H. Lu and Y. Zhang, *Adv. Mater.*, 2006, **18**, 3094.
41. M. H. Nurmawati, R. Renu, P. K. Ajikumar, S. Sindhu, F. C. Cheong, C. H. Sow and S. Valiyaveettil, *Adv. Funct. Mater.*, 2006, **16**, 2340.
42. H. H. Tsai, Z. H. Xu, R. K. Pai, L.Y. Wang, A. M. Dattelbaum, A. P. Shreve, H.-L.Wang and M. Cotlet, *Chem. Mater.*, 2011, **23**, 759.
43. S. U. Pickering, *J. Chem. Soc. Trans.*, 1907, **91**, 2001.
44. V. Vohra, A. Bolognesi, G. Calzaferri and C. Botta, *Langmuir*, 2009, **25**, 12019.
45. A. Boker, Y. Lin, K. Chiapperini, R. Horowitz, M. Thompson, V. Carreon, T. Xu, C. Abetz, H. Skaff, A. D. Dinsmore, T. Emrick and T. P. Russell, *Nat. Mater.*, 2004, **3**, 302.
46. H. Sun, H. L. Li and L. X.Wu, *Polymer*, 2009, **50**, 2113.
47. W. Sun, Z. Shao and J. A. Ji, *Polymer*, 2010, **51**, 4169.
48. X. L. Jiang, X. F. Zhou, Y. Zhang, T. Z. Zhang, Z. R. Guo and N. Gu, *Langmuir*, 2010, **26**, 247.
49. H. J. Zhao, Y. M. Shen, S. Q. Zhang and H. M. Zhang, *Langmuir*, 2009, **25**, 11032.
50. L. Li, Y. W. Zhong, C. Y. Ma, J. Li, C. K. Chen, A. J. Zhang, D. Tang, S. Xie and Z. Ma, *Chem. Mater.*, 2009, **21**, 4977.
51. Y. Sakatani, C. Boissière, D. Grosso, L. Nicole, G. J. A. A. Soler-Illia and C. Sanchez, *Chem. Mater.*, 2008, **20**, 1049.
52. A. Lafuma and D. Quéré, *Nat. Mater.*, 2003, **2**, 457.
53. H. Yabu, M. Takebayashi, M. Tanaka and M. Shimomura, *Langmuir*, 2005, **21**, 3235.
54. P. Escalé, L. Rubatat, C. Derail, M. Save and L. Billon, *Macromol. Rapid Commun.*, 2011, **32**, 1072.
55. Y. Hirai, H. Yabu and M. Shimomura, *Colloids Surf.*, 2008, **313**, 312.
56. Y. Hirai, H. Yabu, Y. Matsuo, K. Ijiro and M. Shimomura, *Chem. Commun.*, 2010, **46**, 2298.
57. L. A. Connal, R. Vestberg, P. A. Gurr, C. J. Hawker and G. G. Qiao, *Langmuir*, 2008, **24**, 556.

58. F. Galeotti, I. Chiusa, L. Morello, S. Giani, D. Breviario, S. Hatz, F. Damin, M. Chiari and A. Bolognesi, *Eur. Polym. J.*, 2009, **45**, 3027.
59. J. H. Kim, M. Seo and S. Y. Kim, *Adv. Mater.*, 2009, **21**, 4130.
60. M. Kojima, T. Nakanishi, Y. Hirai, H. Yabu and M. Shimomura, *Chem. Commun.*, 2010, **46**, 3970.

CHAPTER 6

Nanomaterials with Light Shaping Capabilities: Photonic Crystals

In Chapter **3** we discussed the stick and slip method's capability to induce the formation of spatially-ordered nanoparticle structures. The spatial order represents a phenomenon that occurs in natural processes as well. When referring to spatially ordered nanoparticles obtained by natural processes, we can consider opals as one of the most significant examples. Opals are gemstones[1] with irregular geometrical shapes. The word opal originally comes from Sanskrit (*upala* means 'precious stone'). Later on it was used in Greek and Latin, *opallios* and *opalus*, both meaning 'to see a color change'.[2] Again, Shakespeare referred to opals as 'the miracle [...] and queen of gems'[3,4] due to their fascinating feature to 'manipulate' the light and to show different colors depending on the gem's orientation/position. The property of presenting different colors was initially and tentatively assigned to inclusion of liquids or materials with different refractive indices. Further studies demonstrated that the ability to interact with the light was due to the particular nanostructure. In fact, opals are made of uniform sized amorphous silica spheres in the 0.14–0.30 μm diameter range which are usually arranged in a primitive cubic configuration.[5] The silica spheres are generated from geothermal alkaline water that is saturated with silica simply through cooling or pH change processes. The resultant colloidal solution with an almost monomodal size distribution of particles is then the precursor solution for a close-packed arrangement of the ceramic particles. To date, the time estimated for the formation of 10 mm thick opal is about 5 million years. During such a long natural processing time, billions of silica nanoparticles are arranged in periodic

Water Droplets to Nanotechnology: A Journey Through Self-Assembly
By Plinio Innocenzi, Luca Malfatti and Paolo Falcaro
© P. Innocenzi, L. Malfatti and P. Falcaro 2013
Published by the Royal Society of Chemistry, www.rsc.org

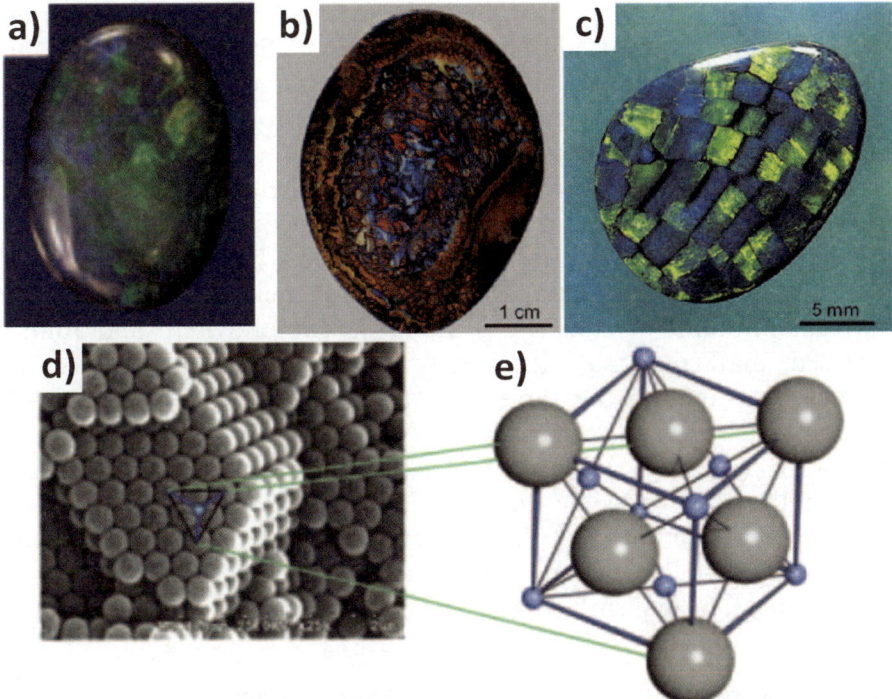

Figure 6.1 Different types of natural opals, such as (a) regular opal, and (b) iron oxide-containing matrix opal, and (c) harlequin opal. (d) Scanning electron microscopy (SEM) image of the opal nanostructure. (e) Arrangement of the particles in the nanostructure lattice (faced-centered cubic arrangement). (Reprinted with permission from F. Marlow *et al.*, *Angew. Chem., Int.* Ed., 2009, **48**, 6212. Copyright 2009, John Wiley & Sons.)

nanostructures. This perfect periodicity ends up macroscopically giving impressive color changes when the stone is moved (see **Figure 6.1**).

One question that should arise from this information is: what is the effect that induces the perception of different colors by using colorless nanoparticles? This effect, called *iridescence* or *goniochromism*, is induced by the interference between the light and the ordered nanostructure. The observed structural color results when light interacts with physical structures featuring a dielectric discontinuity. Such a discontinuity needs to be organized in a spatial scale, which is comparable to the incident light wavelength. Iridescence may also occur over any range of wavelengths (it is not necessarily restricted to the range visible to human eyes).[6] Related effects include changes of other light features such as light polarization and intensity. To date, such a mechanism for controlling the light interaction using nanostructures has been used in biological life since the Cambrian diversification occurred 500 million years ago.[7] Such a smart way to manipulate light is used by plants, insects, fishes and birds. Examples of the ability of nature to engineer nanostructures for light manipulation are shown in the eyes and wings of butterflies.[8] For instance, **Figure 6.2** illustrates the nanostructure in the butterfly where the mechanism of processing light is induced by optimized biological nanostructures.

Synthetic structures presenting a similar effect have also become of interest to physicists, material scientists and chemists who named these materials photonic crystals (PCs). In 1987, Yablonovitch[9] and John[10] disclosed, mathematically, the ability of some ordered structures to interact with light in the same way as semiconductors interact with electrons; such interactions are described by *band theory*.[11] In particular some PCs present the ability of blocking the propagation of certain wavelengths in all possible directions. These PCs are called *complete photonic band gap devices*.[6] The interaction between the light and the ordered dielectric constant variation can be understood using a mathematical formalism[12] that is beyond the scope of this chapter.

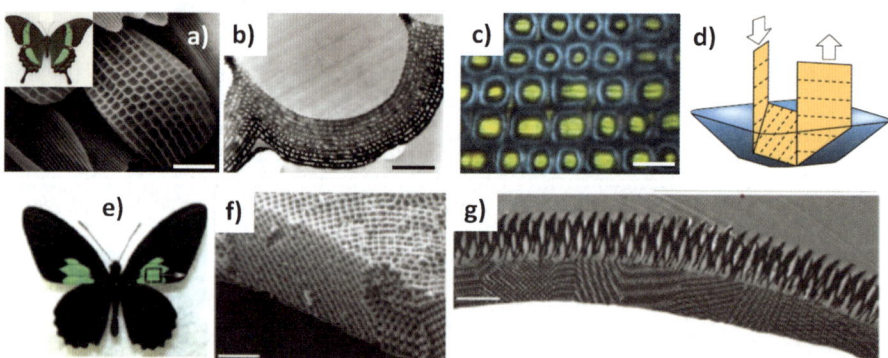

Figure 6.2   Iridescence in butterflies. In the upper row the features of the *Papilo Palinuros* are presented. (a) SEM of an iridescent scale presenting its array of cavities, each with a section that shows curved multilayers (b) measured with the transmission electron microscope (TEM). As shown from the optical microscope image (c), this structure shows two simultaneous structural colors (*i.e.* yellow and blue). The blue ring is created by a double reflection from opposite and perpendicular cavity sides (d). The schematic also illustrate the light polarization change. In the lower row the features of *Parides Sesostris* (e) are presented. (f) The SEM image of the photonic crystal of the green part of the wing is shown. (g) TEM image showing a 50 nm section through the scale. Scale bars, (a) 15 mm; (b) 1 mm; (c) 6 mm; (f) 1.2 mm; (g) 2.5 mm. (Adapted with permission from: P. Vukusic and J.R. Sambles, *Nature,* 2003, **424**, 652, Copyright 2003, Macmillan Publishers Ltd.)

Indeed, we will mostly focus on fabricating the crystal formation through self-assembly evaporative processes.

6.1 WHICH TYPE OF SELF-ASSEMBLED PHOTONIC CRYSTALS?

In general, three different types of photonic crystals can be prepared. 1D-, 2D- and 3D-photonic crystals are classified depending on the ordered spatial repetition of different dielectric materials.

Alternating layers of materials can easily generate a photonic crystal with different dielectric constants in one direction (1D-photonic crystal). The dielectric stack takes advantage of Bragg's law for the interference of reflected light from the different layers.[12] This material, technically called a multilayer film, can act as a mirror for incident light with a specific wavelength; it can localize light modes if there are any defects in the multilayer structure.[12]

A 2D-photonic crystal is a periodic structure along a plane. The ordered structure is repeated along two of its axes (*e.g.* x and y) while it is homogeneous along the third axis direction. A typical example is a structure of alternated tall square pillars on a surface (**Figure 6.3**). For proper values of the column spacing, this crystal can have a photonic band gap in the plane (xy). Within this gap the incident light is reflected. Unlike the multilayer film, this 2D-photonic crystal can prevent light propagation along any direction within the xy plane.

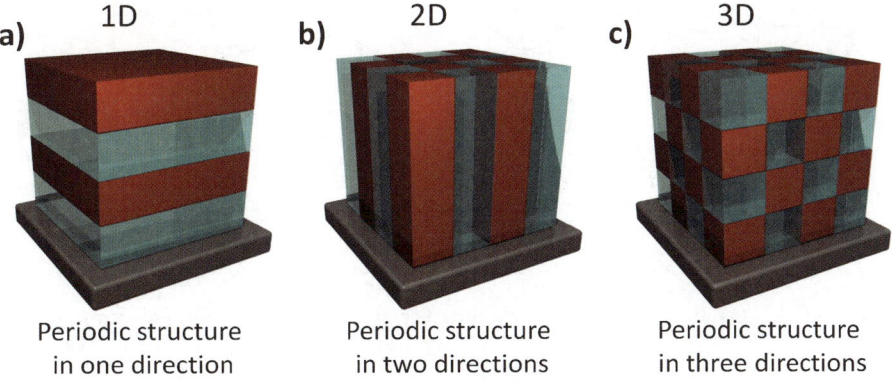

Figure 6.3 Simple examples of 1-, 2-, and 3D photonic crystals (a), (b) and (c), respectively. The red and light blue-transparent materials represent materials with different dielectric constants. The main feature of a photonic crystal is the periodicity of a dielectric material along one or more axes.

By increasing the complexity of the structure (repetition of the structure along the *z* axis) a 3D-photonic crystal can be obtained. An example of alternating boxes along the three axes is proposed in **Figure 6.3(c)**. Although there are an infinite number of possible geometries for a 3D-photonic crystal, the technological interest is mainly related to those geometries that promote the existence of photonic band gaps. Under these conditions the additional capability to localize light in all three dimensions can be achieved.[12] However, not all the geometries allow for an efficient interaction along any direction. For instance, in a face centred cubic (FCC) arrangement of spherical objects with a high dielectric constant (ε) in air the photonic modes are overlapped; this does not allow for a full band gap (mode separation) as presented in **Figure 6.4**. However, more complex structures would offer a complete band gap. In particular, a reverse-diamond-like arrangement (spheres made by air surrounded by a material with high ε, $\varepsilon = 13$) allows for the formation of a complete band gap (net separation between the photonic modes) as presented in **Figure 6.4(b)**. Although the two different structures (FCC and reverse diamond) are made by using materials with $\varepsilon = 13$ and air ($\varepsilon = 1$) it is clear how the spatial arrangement of the different dielectric materials plays a crucial role in the properties of the device.

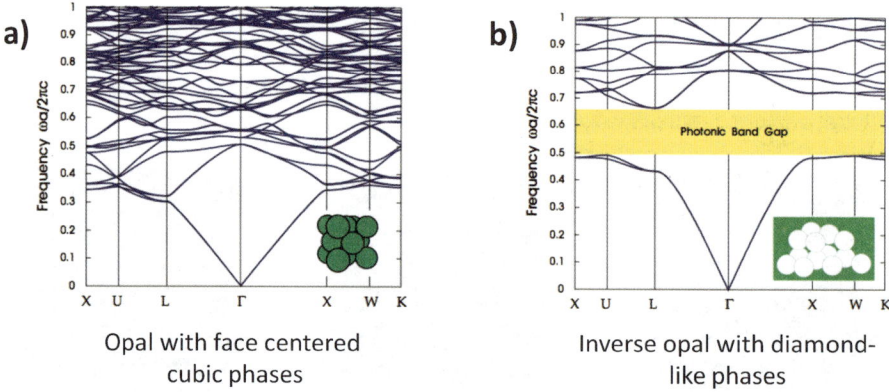

Opal with face centered cubic phases

Inverse opal with diamond-like phases

Figure 6.4   (a) Photonic band structure for the lowest-frequency electromagnetic modes of a lattice with FCC symmetry. The inset presents the symmetry of the dielectric spheres ($\varepsilon = 13$) in air. The band diagram shows the absence of a complete photonic band gap. (b) Photonic band structure for the lowest bands of a diamond lattice of air spheres in the same dielectric ($\varepsilon = 13$) material. A complete photonic band gap induced by this symmetry (inset) is shown with a yellow color. (Reprinted with permission from J. D. Joannopoulos *et al.*, *Photonic Crystals, Moulding the flow of light*, Princeton University Press, New Jersey, 2008.)

Due to their superior ability to control light in space we will consider mainly photonic crystals with a 3D arrangement. In particular, we will focus on the fine control of positioning the material in a 3D lattice.

6.2　HOW TO FABRICATE PHOTONIC CRYSTALS

Photonic crystals are currently prepared by using micro- and nano-fabrication protocols. In particular, widely-used procedures for the fabrication of such optical band-gap devices include lithographic protocols with radiation ranging from ultra-violet (UV) light[13] to hard X-ray radiation.[14] The radiation is used to induce a spatially controlled chemical change in a photosensitive material (*e.g.* polymerization). Further removal of a selected region (either the exposed or the unexposed regions) enables the formation of a spatially controlled refractive index modulation. These processes are typically called top-down because 2D and 3D micro- or nano-metric periodic structures can be fabricated from a macroscopic bulk material (monolith or film).

In general, top-down fabrication methods allow for a fine control of the features within complicated patterns. However, such lithographic protocols are either cost effective and/or time expensive. In particular, the higher the pattern resolution required in a large area the more expensive the lithographic process; this is the case of the patterning process on the nanoscale.

However, the ability to take inspiration from natural processes and to translate natural methods for industrial purposes would offer a cheap and fast route for device fabrication. In particular, the ability to take advantage of self-organizing materials for the preparation of photonic structures would offer a competitive route to achieve ordered nano- and micro-structured materials. The control of the process parameters would be enough to trigger the self-organization of periodic structures on large surfaces. The advantage of the proposed method would potentially offer a straightforward route for massive fabrication. Therefore, understanding self-assembly processes is of paramount importance not only to satisfy human curiosity but also for technological applications.[15]

6.3　SELF-ASSEMBLED PHOTONIC CRYSTALS

There are two main different types of photonic crystal preparations using self-assembly: block copolymer assembly and colloidal assembly.

Block copolymer assembly involves two or more chemically different polymer chains (*blocks*) linked together. Under the proper conditions, the system evolves towards the minimum free energy configuration, which corresponds to a local segregation (typically 10–100 nm phase segregation with a sharp interface).[16,17] In the case of the simplest copolymer (AB diblock copolymer, where A and B are two different block polymer chains), at least four different spatial arrangements can be obtained. The morphologies include spherical, cylindrical, gyroid and lamellar structures.[18] By using a third block at least 12 different and more complex arrangements can be obtained.[19]

Colloidal assembly is obtained from a colloidal solution where a continuum phase, typically a solvent (*e.g.* water or alcohol), contains monodispersed sub-micrometric or micrometric particles as a dispersed phase. The colloidal assembly occurs if the colloidal particles spontaneously arrange into an ordered crystalline lattice.

In general, the self-assembly of colloidal crystals can be successfully achieved using different deposition methods including gravity sedimentation, vertical deposition, electrophoresis, spin coating, and crystallization in physically confined cells.[20] However, among all the deposition methods the most popular one is vertical deposition as proposed by Colvin *et al.*[21] in 1999. With this procedure, opaline coatings with monodispersed particles having diameters below 400 nm are easily prepared. The method, based on solvent evaporation, will be discussed in the next section. Interestingly, in 2001 the method was extended to bigger spherical particles that allows for the preparation of photonic crystals using monodispersed particles up to 1000 nm.[22] Such an extension would potentially allow for the integration of the photonic crystals within the optical fibres for telecommunications because of the band gap at 1.55 μm.[23] In fact, such data transmission, depending on the generation of the considered optical fibre, uses wavelengths in the 1.3–1.6 μm range. In **Figure 6.5** a cross-section of a coating that presents an ordered arrangement of micrometric particles is shown. The

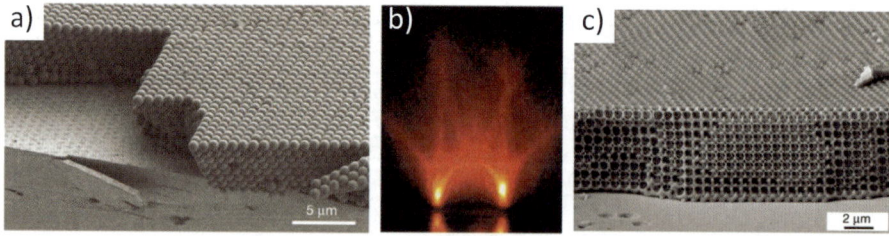

Figure 6.5   (a) SEM cross-section of an opaline film. (b) Corresponding optical diffraction pattern. (c) Inverse opal obtained using nanoparticles as templates. (Reprinted by permission from: Y.A. Vlasov *et al.*, *Nature*. 2001, **414**, 289. Copyright 2001, Macmillan Publishers Ltd.)

interaction with light and the cross-section of the inverse opal obtained from the opaline film are also presented.

6.4 BAND GAP, OPALS AND INVERSE OPALS

As we have previously disclosed, the chance to easily manipulate light drives a strong interest in photonic crystals as processing devices. In fact, if such a device was made by materials absorbing a minimal amount of light while adequate different dielectric constants are repeated within the crystal, then the refractions and reflections of light from all of the various interfaces could produce many of the same phenomena for photons (light modes) that the atomic potential produces for electrons. Therefore, the chance to efficiently employ photonic crystals for light processing is strongly related to the way the material interacts with the photon propagation.[12] In other words, the behavior of photonic crystals depends on their band gap features. To date, a complete (or full) band gap is defined as one that can extend over the entire *Brillouin zone* (the primitive cell in reciprocal space[11]) in the photonic band gap structure.[24] The properties of full band gap optical devices would offer the chance to control light propagation in all possible directions. As an example, electronics relies on semiconductors that offer a complete band gap between the conduction band and the valence band. In a similar way, the ability to design complete optical devices with a customized complete band-gap would therefore give photonic crystals a real chance of being used for light processing.

Unfortunately, the spherical symmetry of the spherical particles inhibits the formation of a complete band gap that would be a crucial feature of technological interest. Recent calculations also suggested that the symmetry limitation could be overcome by using colloidal spheres made of materials with particular properties (*i.e.* large magnetic susceptibilities or intensity-dependent refractive indices).[24] A different route for the preparation of self-assembled materials useful for photonic application is the self assembly of non-spherical objects. However, the use of non-spherical objects makes the self-assembly process less efficient. Current research is now facing this scientific challenge.

However, inverse opals made from spherical objects offer the chance to prepare complete photonic band gap materials.[25] For this reason, bare opals are often used as sacrificial scaffolds (template) for further processing including infiltration and removal of the template. A schematic of the procedure is presented in **Figure 6.6**.

Although different steps are needed for the preparation of a self-assembled full band gap photonic crystal, the initial opal formation is a

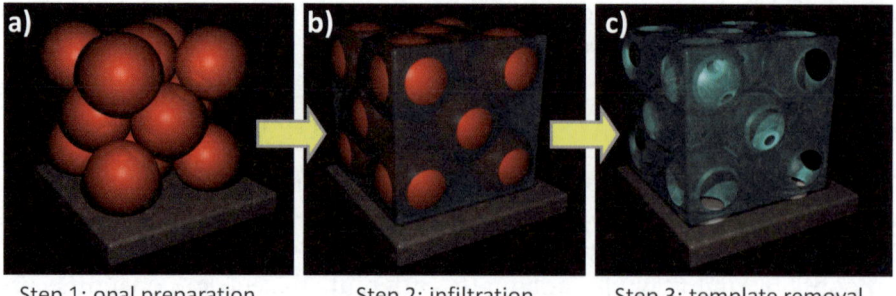

Step 1: opal preparation   Step 2: infiltration   Step 3: template removal

Figure 6.6   Inverse opal fabrication steps involving the preparation of the opal as
template (a), the infiltration (b) and the template removal (c).

crucial milestone. In fact, the crystal quality and crystal lattice
symmetry are among the most important parameters for the optical
properties of the colloidal crystals.

6.5   SELF-ASSEMBLY OF SPHERICAL PARTICLES DRIVEN BY SOLVENT EVAPORATION

At the beginning of 1990, important studies were performed to prepare
2D systems (*i.e.* colloidal monolayers).[26] However, it was only in 1995
that a 3D arrangement of particles (artificial opal) for an optical
application was prepared using a self-assembly technique.[27]

For an understanding of the photonic crystal formation using self-
assembly from colloidal solution, we will initially refer to the
mechanism for the formation of a colloidal monolayer. Such a 2D
system relies on lateral capillary forces that become effective during
solvent evaporation.[23] In fact, when the spheres protrude from the
solvent layer the capillary force acts by pushing the particles in a dense
2D packing arrangement. A schematic visualising the step processes is
presented in **Figure 6.7**.

For vertical deposition, where a substrate is immersed in a colloidal
solution and extracted at a constant withdrawal rate (dip-coating
method), evaporation of the solvent and deposition of the particles
by lateral capillary forces are allowed only at the triple point of
the suspension−substrate−air interface. A large-scale polycrystalline
monolayer of a colloidal array with domains ranging from a few tens of
nanometers to a few micrometers can be obtained with this method.

This mechanism has been used for the preparation of more complex
photonic crystals such as 3D particle lattice, however a lack of quality
in the final crystal has been observed.

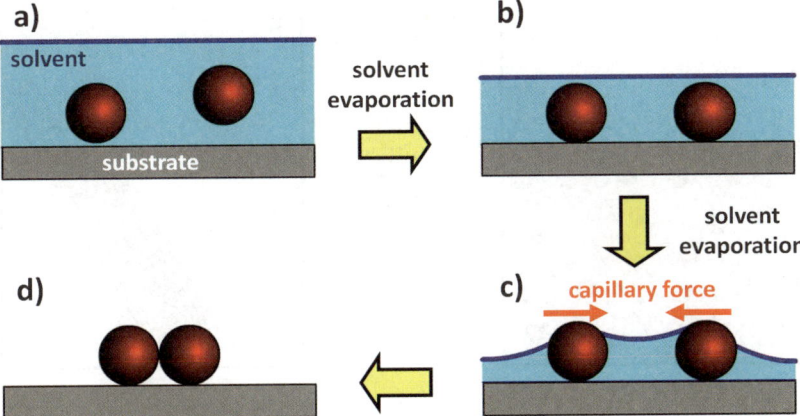

Figure 6.7 Schematic of the capillary force acting between particles during the evaporation process.

A procedure enabling the fabrication of high quality 3D colloidal crystals was proposed by Colvin *et al.*[21] This protocol involves a colloidal solution with about 1% of colloidal silica spheres. A substrate is immersed in such a suspension of ceramic spheres in a solvent (*e.g.* water). Under the correct conditions, evaporation of the solvent induces the 3D ordered packing of the particles starting from the surface of the colloidal solution. A schematic of the procedure is presented in **Figure 6.8**. The thickness of the fabricated coating is generally homogeneous and, after the complete solvent removal, the spheres adhere to each other.[23] The resulting film is robust enough to be handled and re-immersed in a new solution without affecting the film quality. While, for silica nanoparticles up to 400 nm, water can be successfully used as a solvent, for bigger ceramic spheres the deposition method works whereas particular conditions are used. In fact a more volatile solvent is required (*e.g.* ethanol) and a very narrow concentration range is permitted.

Despite the remarkable result offered by this approach, the mechanism of packing is not yet understood. A potential explanation is provided by the convective self-assembly theory[23] proposed in Chapter **4**.

To date, under a thermodynamic equilibrium, it has been theoretically demonstrated that hard spheres preferentially self-assemble in a closed-packed FCC arrangement rather than the hexagonal one (HCP), which is slightly less stable.[28]

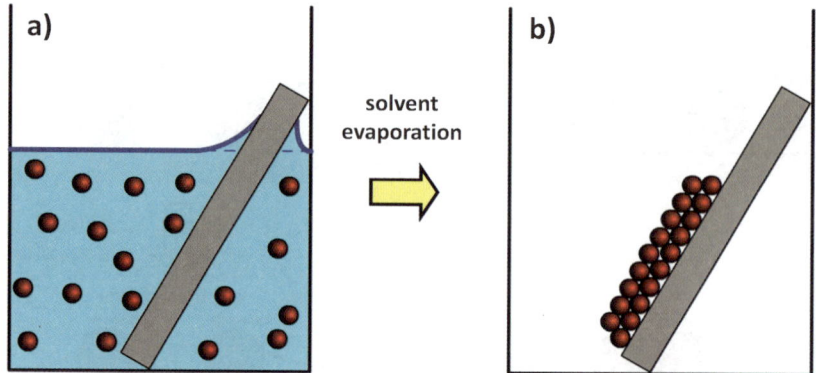

Figure 6.8 Formation of an opaline coating using a tilted substrate during the evaporation of the solvent. (Reprinted with permission from D. J. Norris *et al.*, *Adv. Mater.*, 2004, **16**, 1393. Copyright 2004, John Wiley & Sons.)

6.6 SPHERICAL PARTICLES COMPOSITION AND DEPOSITION PARAMETERS IN EVAPORATIVE SELF-ASSEMBLY

Among the different monodispersed spherical particles used for the preparation of colloidal crystals, silica (SiO_2) and polymers are the materials that are most widely used. Polymeric beads are mainly prepared using polystyrene (PS) or polymethyl methacrylate (PMMA) due to well-established preparation methods. In general, the choice of

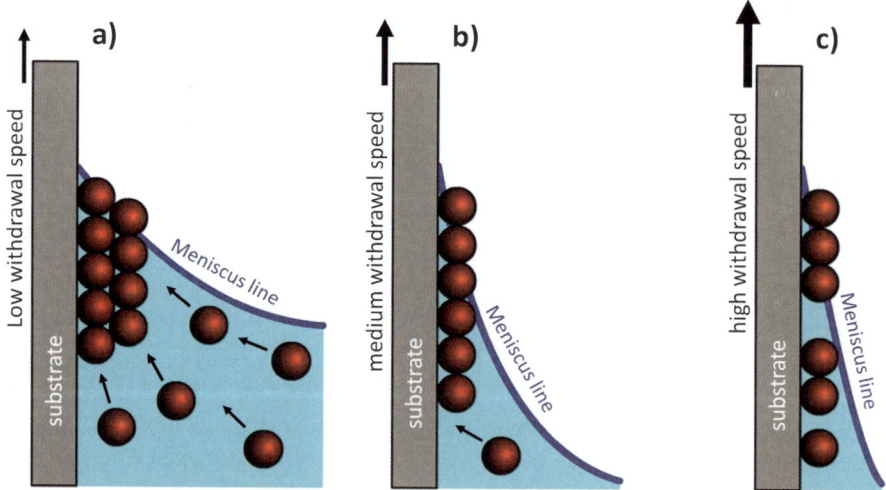

Figure 6.9 Schematic of the meniscus shape at increasing withdrawal speed. (Reprinted from J. Zhang *et al.*, *Curr. Opin. Colloid Interface Sci.*, 2009, **14**, 103. Copyright 2009, with permission from Elsevier.)

polymeric particles *versus* ceramic beads offers the chance to easily tune the size and the composition following inexpensive preparation methods. If the opal is used as a sacrificial template for the fabrication of an inverse opal, the polymer can be easily removed using either solvent or thermal treatment.[29]

Polystyrene (latex) bead films have been investigated over many decades and a fundamental understanding has been achieved. The formation of colloidal crystals based on fused polyhedra evolves along three main steps:[23]

(1) consolidation: while the solvent fills the interspaces between the particles, each sphere locks in contact with other particles

(2) compaction: the particles efficiently pack against each others with a subsequent pore space reduction

(3) coalescence: the particles interdiffuse connecting the spheres together. This step provides a continuous polymer coating.

Although these steps can be visualized conceptually (separately) they can occur simultaneously.

Moreover, in case of a lack of control during the self-assembly process, a disordered arrangement of the spherical particles can be obtained. However, if the deposition parameters are not properly set, defects in the lattice (*e.g.* vacancies,[30] stacking faults[30] and macroscopic cracks) can compromise the photonic crystal quality and the functional properties as well.

As a case of study, we can consider vertical deposition; as already mentioned, this is the most popular methods for the preparation of photonic crystals through self-assembly. In this case, the parameters that need to be controlled are mainly variables that can affect the evaporation rate including the type of solvent, the air pressure, and the humidity. Here we will discuss in detail the effects induced by the listed parameters.

The solvent plays a crucial role in the colloidal crystal self-assembly; it drastically affects the ordering of the microspheres, the thickness of the film and the crack formation. In our previous discussion on the material used for the spherical particle preparation, we presented how switching from water to ethanol has been the key to self-assembled silica particles with sizes bigger than 400 nm. In this context, a systematic investigation of varying the composition of the ethanol−water mixture highlighted the importance of the surface tension, viscosity and volatility of the solvent.[31]

Once the solvent is chosen, the air pressure is a variable that can affect the evaporation rate. Studies related to different pressure, including vacuum, have shown a strong influence on the crystal quality and film morphology.[20]

The humidity is another important variable since it affects the capillary force between the particles. It has been demonstrated, at a fixed temperature, how a high relative humidity leads to the formation of higher quality crystals. The explanation is that the humidity simultaneously helps the correct arrangement of the particles in an ordered lattice and decreases the probability of defect occurring.

If the process used for the deposition involves the withdrawal of the substrate from a colloidal solution with fixed concentration, the correct choice of the lift-up rate is another important parameter (Figure 6.9). At low withdrawal speed multi-layered coatings can be prepared due to the high number of particles entering the suspension−substrate−air interfacial region. However, at high withdrawal speeds, the number of particles entering in the same interfacial region might be not sufficient to provide a continuous layer.

REFERENCES

1. The National Opal Collection, Australia, http://www.nationalopal.com/
2. F. Marlow, Muldarisnur, P. Sharifi, R. Brinkmann and C. Mendive, *Angew. Chem., Int. Ed.*, 2009, **48**, 6212.
3. A.W. Eckert, *The World of Opals*, Wiley, New York, 1997.
4. K. Nassau, *Gems Made by Man*, Chilton Book Company, 1980.
5. http://www.resources.nsw.gov.au__/data/assets/pdf_file/0011/398099/QN136.pdf
6. L. Poladian, S. Wickham, K. Lee and M. C. J. Large, *J. R. Soc. Interface*, 2009, **6**, S233.
7. P. Vukusic and J. R. Sambles, *Nature*, 2003, **424**, 652.
8. S. Lou, X. Guo, T. Fan and D. Zhang, *Energy Environ. Sci.*, 2012, **5**, 9195.
9. E. Yablonovitch, *Phys. Rev. Lett.*, 1987, **58**, 2059.
10. S. John, *Phys. Rev. Lett.*, 1987, **58**, 2486.
11. C. Kittel, *Introduction to Solid State Physics*, John Wiley & Sons, Inc., New York, 1996.
12. J. D. Joannopoulos, S. G. Johnson, J. N. Winn and R. D. Meade, *Photonic Crystals, Moulding the flow of light*, Princeton University Press, New Jersey, 2008.
13. J. H. Moon, J. Ford and S. Yang, *Polym. Adv. Technol.*, 2006, **17**, 83.
14. P. Innocenzi, L. Malfatti and P. Falcaro, *Soft Matter*, 2012, **8**, 3722.
15. J. F. Galisteo-López, M. Ibisate, R. Sapienza, L. S. Froufe-Pérez, A. Blanco and C. López, *Adv. Mater.*, 2011, **23**, 30.
16. A. M. Urbans, M. Maldovan, P. Derege and E. L. Thomas, *Adv. Mater.*, 2002, **14**, 1850.
17. J. H. Moon and S. Yang, *Chem. Rev.*, 2010, **110**, 547.

18. F. S. Bates and G. H. Fredikson, *Ann. Rev. Phys. Chem.*, 1990, **41**, 525.

19. W. Zheng and Z.-G. Wang, *Macromolecules*, 1995, **28**, 7215.

20. J. Zhang, Z. Sun and B. Yang, *Curr. Opin. Colloid Interface Sci.*, 2009, **14**, 103.

21. P. Jiang, J. F. Bertone, K. S. Hwang and V. L. Colvin, *Chem. Mater.*, 1999, **11**, 2132.

22. Y. A. Vlasov, X. Z. Bo, J. C. Sturm and D. J. Norris, *Nature*, 2001, **414**, 289.

23. D. J. Norris, E. G. Arlinghaus, L. Meng, R. Heiny and L. E. Scriven, *Adv. Mater.*, 2004, **16**, 1393.

24. Y. Xia, B. Gates and Z.-Y. Li, *Adv. Mater.*, 2001, **13**, 409.

25. S. A. Rinne, F. García-Santamaría and P. V. Braun, *Nat. Photonics*, 2008, **2**, 52.

26. H. D. Denkov, O. D. Velev, P. A. Kralkevsky, I. B. Ivanov, H. Yoshimura and K. Nagayama, *Langmuir*, 1992, **8**, 3183.

27. V. N. Astranov, V. N. Bogomolov, A. A. Kaplyanskii, A. V. Prokofiev, L. A. Samolovich, S. M. Samolovich and Y. A. Vlasov, *Nuovo Cimento*, 1995, **17D**, 1349.

28. L. V. Woodcock, *Nature*, 1997, **385**, 141.

29. A. Stein, F. Li and N. R. Denny, *Chem. Mater.*, 2008, **20**, 649.

30. W. D. Callister, *Materials Science and Engineering: An Introduction*, John Wiley & Sons, UK, 2007.

31. H. L. Li and F. Marlow, *Chem. Mater.*, 2008, **20**, 649.

CHAPTER 7

Superlattices and Quasicrystals

The self-assembly of photonic crystals is a process that involves the self-organization of preformed objects on the micron scale; this allows us to obtain materials whose interaction with light has a striking effect. Evaporation-driven self-organization into 3D structures works quite well for micron objects, but what happens if we move to colloidal dimensions? Here things become a little more complicated because colloidal forces start to take effect and the stability of the suspension has to be controlled to avoid aggregation during self-assembly. However, colloidal nanocrystals have size-dependent properties that are not observed in their bulk counterpart and offer a unique opportunity for developing materials with new electronic, photonic and magnetic properties. If they form organized 2D and 3D arrays, these can be imagined as a kind of 'crystal of crystals' whose properties are controlled at the nanoscale of the building units and the regular organization of the structure. In this chapter we will see two examples of such nanomaterials that are formed upon evaporation of colloidal nanocrystals: superlattices and quasicrystals.

7.1 COLLOIDAL NANOCRYSTALS AND SUPERLATTICES

The first step for achieving ordered structures from colloidal particles is to have the right building units and the first question we have to face is how big should these particles be? To be able to fully exploit the size-dependent properties we should remain within the 1–50 nm range because within these dimensions the nanocrystals behave like quantum objects. However, size is not the only property that we have to factor in; several other properties such as chemical composition, shape, and surface structure, will affect the properties of the nanoparticles and their

Water Droplets to Nanotechnology: A Journey Through Self-Assembly
By Plinio Innocenzi, Luca Malfatti and Paolo Falcaro
© P. Innocenzi, L. Malfatti and P. Falcaro 2013
Published by the Royal Society of Chemistry, www.rsc.org

stability in solution.[1] The synthesis of colloidal nanocrystals has reached a high level of sophistication and a surprisingly large number of particles of different nanosize, shape and composition have been produced so far.[2–4] These nanoparticles form a nice toolbox for obtaining new materials with tailored chemical and physical properties.[5] The capability of getting monodispersed nanoparticles that are able to form stable suspensions is the key point. If a precise size distribution is achieved the nanocrystals exhibit, under the proper conditions, a surprising capability to self-assemble into ordered 3D structures, which are indicated as superlattices. There are, however, some striking differences between these 'crystals of crystals' and ordinary solid-state crystalline materials, which make the superlattices much more interesting. The first difference is a structural difference, which is due to the binding energy of the building units, atoms in one case and nanocrystals in the other one. The binding energy is given by quantum mechanical interactions between atomic orbitals of close atoms in ordinary crystals whilst only secondary bonds bind neighboring particles in a superlattice. Even if the particles in a superlattice held together by weak interactions, such as van der Waals attractions, they can exhibit surprisingly good mechanical stability and strength.[6,7] However, the great attractiveness of nanocrystal superlattices is not really related to their mechanical response but to the new collective properties that arise from an ordered array of single nanocrystals. The superlattices' properties can be engineered at a very fine level by tuning the size, shape, composition of the nanocrystals and the interparticle distance.

7.2 AGAIN, STARTING FROM A DROPLET

We started this book with a chapter dedicated to the coffee-stain effect. This was not a casual choice (we had a clear intention from the very beginning that most of the phenomena that are connected with evaporation and self-assembly are far from equilibrium, and system instability represents quite a general case). What makes for an interesting challenge is trying to get order in situations that generally favor disorder and this can be achieved if the forces that work during evaporation are understood and controlled. In the case of a solvent droplet containing particles from the nano to micron range, the liquid–air interface is of paramount importance and the surface energy dictates, to a large extent, the final disposition of the particles on the surface after evaporation. In the case of nano-objects, the van der Waals interaction is the attractive force that drives self-assembly, while the nanoparticle surface ligands allow for the creation of a stable structure.[8] In general, the evaporation of a droplet containing nanocrystals could leave several types of different patterns behind, which range from 2D and 3D superlattices to fractal-like aggregates and percolated structures. The degree of order can vary a lot and it is relatively easier to obtain domains of different sizes; whereas forming a superlattice that displays long-range order is much more difficult.[9] How is it possible to form an ordered monolayer of nanoparticles that extends over a long range, at least for some millimeters? The answer is connected with the surface, *i.e.* nanoparticle and droplet, and the kinetics of the process. If we can control these two key features order arises and long-range superlattice structures form.

To follow the process in more detail we can use gold nanoparticles whose surface is modified through dodecanethiol as a model. Dodecanethiol is a molecule that can be easily linked to the gold surface. After functionalization a capping shell of dodecanethiol covers the gold nanoparticles. This layer is not perfect as some voids can form, which favors the interdigitation of the alkane chains, which in turn increases the van der Waals interaction. Introducing a slight excess of thiol in the solution is an important procedure because not only will it help to interlock the particles in the lattice but also having an excess of a less volatile molecule, such as dodecanethiol, will slow down the liquid droplet evaporation. This means that the kinetics of evaporation, as we have previously underlined, is a key point in the process. This importance is confirmed by the observation that by changing the solvent evaporation rate, both 2D and 3D nanocrystal superlattices can be obtained from the same colloidal solution.[10] There is also a small surprise behind the crystallization of nanoparticles in an evaporating droplet because it happens through a nucleation and growth mechanism at the liquid–air interface (**Figure 7.1**). 2D gold nanocrystal superlattices form at the

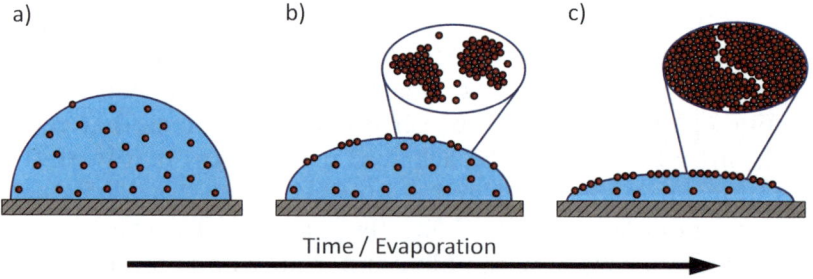

Figure 7.1   Schematic diagram of the self-assembly process during the early stages of drying (not to scale), showing how a quickly receding liquid–air interface captures nanocrystals. (Reprinted with permission from: T. P. Bigioni *et al.*, *Nat. Mater.*, 2006, **5**, 265. Copyright 2006, Macmillan Publishers Ltd.)

liquid−air interface if the evaporation is fast enough for inducing nanocrystals to accumulate at this interface. However, if the evaporation is much slower the nanocrystals diffuse away from the interface and tend to form 3D superlattices inside the droplet. To form organized superlattices through self-assembly at the interface of an evaporating droplet, fast evaporation of the solvent to segregate the nanoparticles close to the liquid–air interface is necessary. An attractive interaction between the particles and the liquid–air interface is necessary to localize them on the interface.[11] This simple strategy can be understood successfully from **Figure 7.2**, which shows the perfection of a monolayer of a gold nanocrystal superlattice reached after the evaporation of a toluene droplet.

The method is quite versatile as it works with different types of nanoparticles and ligands.[12] It should be noted that even if we limit our cases to the evaporation-driven self-assembly of nanocrystals, this is not the only route available and several other strategies have been followed to obtain a large variety of superlattices, such as the Langmuir−Blodgett technique,[13] layer-by-layer deposition,[14] and cross-linking of nanoparticles using templating molecules.[15]

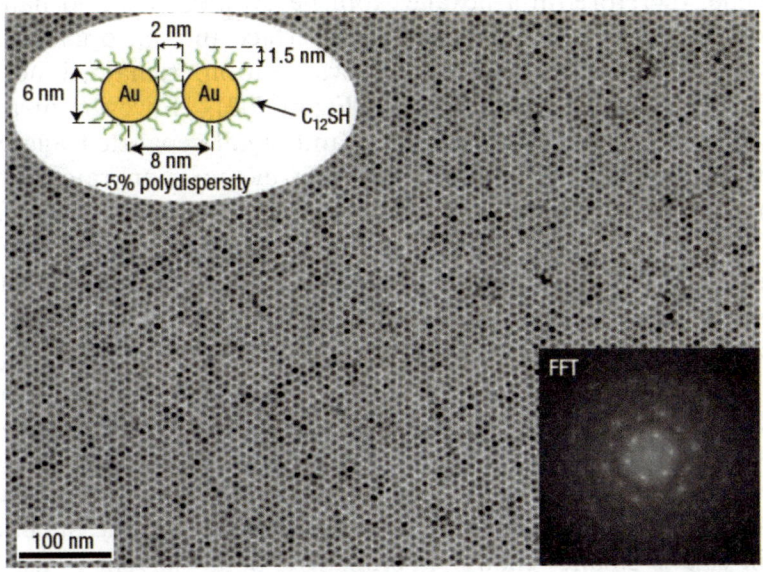

Figure 7.2 Micrograph of a typical monolayer produced by drop-casting 10 µL of a solution of dodecanethiol-ligated 6 nm gold nanocrystals onto a 3 mm × 4 mm substrate. The upper left inset schematically shows the arrangement of two neighboring nanocrystals in the monolayer. The lower right inset is a fast Fourier transform of the image. (Reprinted with permission from: T. P. Bigioni *et al.*, *Nat. Mater.*, 2006, **5**, 265. Copyright 2006, Macmillan Publishers Ltd.)

7.3 EXTENDING THE SUPERLATTICE DOMAINS: CHARGED GOLD NANOPARTICLES IN NON-POLAR SOLVENTS

Controlling the extension of the superlattice array is a fundamental issue especially for practical applications whilst, in general, with nanoparticles of a few nanometers the 2D single crystalline superlattice remains below a few μm. Increasing the superlattice domain size using self-assembly in an evaporating droplet is possible by taking advantage of the liquid−air interface.[16] We have seen that self-assembly in a solvent droplet, such as toluene, requires an excess of thiols in order to stabilize the evaporation rate. This strategy for achieving order is focused on controlling the processing parameters, such as particle concentration, evaporation kinetics and ligand content. The perspective could be changed a little bit and instead of focusing on processing conditions we could control, at first, the properties of the nanoparticles.[17,18] This is the route that has been followed by Eah *et al.* They synthesized colloidal gold nanoparticles that were negatively charged in non-polar solvents and were coated with organic hydrophobic thiol ligands.[19] These colloidal particles were well dispersed in hexane but not in toluene, therefore, in a hexane−toluene droplet, the gold nanoparticles quickly floated to the air−liquid interface and self-organized into a 2D monolayer film, which after evaporation, remained homogeneously deposited on the substrate (**Figure 7.3**). However, in a pure hexane droplet, no film was formed and small separated aggregates were left on the substrates. The different evaporation rates between toluene and hexane, with hexane evaporating around four times faster, drives the particles to the liquid−air interface. The negative charge of the nanoparticles is essential to give the difference in solubility between hexane and toluene, which allows the particles to float at the droplet surface during evaporation. Nanoparticle size is also important, and only nanoparticles with a diameter larger than 4 nm allow the formation of a monolayer both at the liquid−air interface and on the substrate, with almost no limits in size, upon evaporation.

The deposition of a homogeneous monolayer of gold nanoparticles is achieved without using an excess of thiol; however, the superlattice domain size is smaller than 1 micron. If a polar solvent is used (*i.e.* water) a similar strategy can be employed. In this case a hexane droplet containing gold nanoparticles is dropped onto a larger toluene droplet, which has been previously poured on a silicon substrate. Immediately afterwards the nanoparticles will migrate and float to the air–toluene interface. This droplet is then pushed to surround a nearby water droplet and finally a monolayer film of the nanoparticle is deposited on

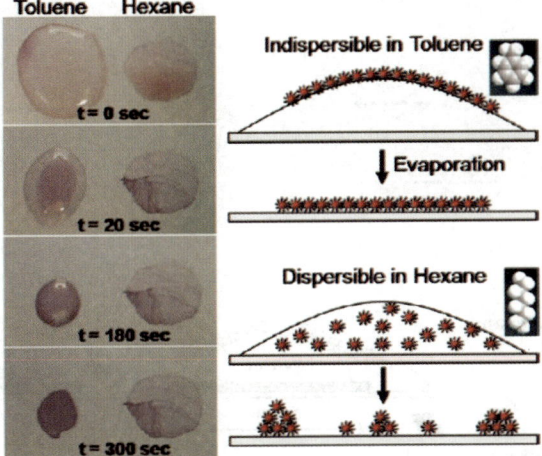

Figure 7.3 When a hexane droplet containing charged gold nanoparticles is mixed with a larger toluene droplet; nanoparticles immediately float to the air–liquid interface and form a monolayer film. After evaporation of the solvent molecules, the monolayer of close packed nanoparticles can be deposited on any substrate. Charged gold nanoparticles in a hexane droplet do not form a monolayer film. (Adapted with permission from M. N. Martin *et al.*, *Langmuir*, 2010, **26**, 7410. Copyright 2010, American Chemical Society.)

the substrate after the evaporation of toluene (first) and water (second) (**Figure 7.4**). The surprise is that at the end of the process the 2D gold superlattices exhibit domains larger than 20 µm and are obtained without an excess of ligands.

7.4   NOT ONLY SPHERES

Obtaining superlattices through self-assembly requires building blocks of well-defined shapes and dimensions and, therefore, not only spherical nano-objects have to be employed. Several oxides and metals can be synthesized in different shapes, and in particular we may consider the case of iron oxide that is able to form superparamagnetic nanocubes of maghemite. These building units have the advantage that, beside the geometrical cubic shape, they show a response to a magnetic field that can be used to modulate the self-assembly process.[20] The colloidal precursor solution contains toluene, maghemite nanocubes, and low volatile additives like 1-octadecene and oleic acid, which contribute to the adjustment of the viscosity and promotes the crystals' rearrangement into a close-packed organization. If the solution is deposited as a droplet, the application during the evaporation stage of a weak

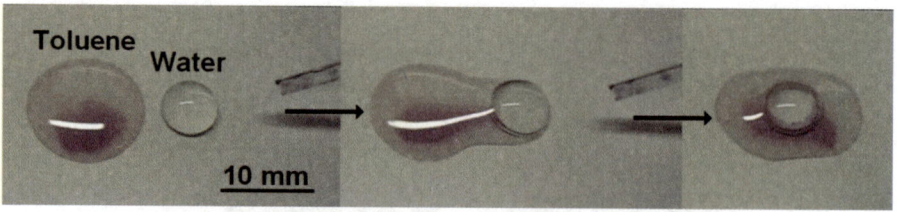

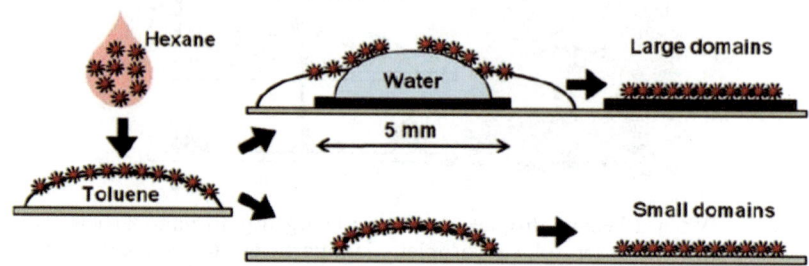

Figure 7.4   Gold nanoparticles of red-purple color and in a 2D liquid-like state at the air–toluene interface are transferred at the air–water interface where they are of blue-purple color and in a 2D solid-like state (images taken from a real-time movie) (top). Very large 2D superlattice domains are formed during transfer of a nanoparticle monolayer film from the air–toluene interface to the air–water interface of toluene and water droplets (bottom). (Reprinted with permission from ref. 18.)

magnetic field will induce a dipolar attraction during the initial stage of the droplet drying that favors the formation of highly ordered structures and almost defect-free superlattices. The superlattices of maghemite nanocubes and the scheme of the self-assembly process are shown in **Figure 7.5**. A permanent magnet, with the main field axis perpendicular to the substrate surface, is slowly moved from one position to another (from 1 to 2) and then moved in the orthogonal direction. The application of the magnetic field favors the formation of a nanocrystal nucleus that drives the aggregation and growth of highly oriented crystals of large dimensions. The final result is a superlattice of superparamagnetic crystals with dimensions up to 10 μm, which contains at least 10^6 nanocubes.

7.5   NOT ONLY DROPLETS

The previous examples of superlattices use the evaporation of a colloidal droplet as the processing method for achieving self-assembled, highly ordered structures. The self-organization can even be 'pushed' by

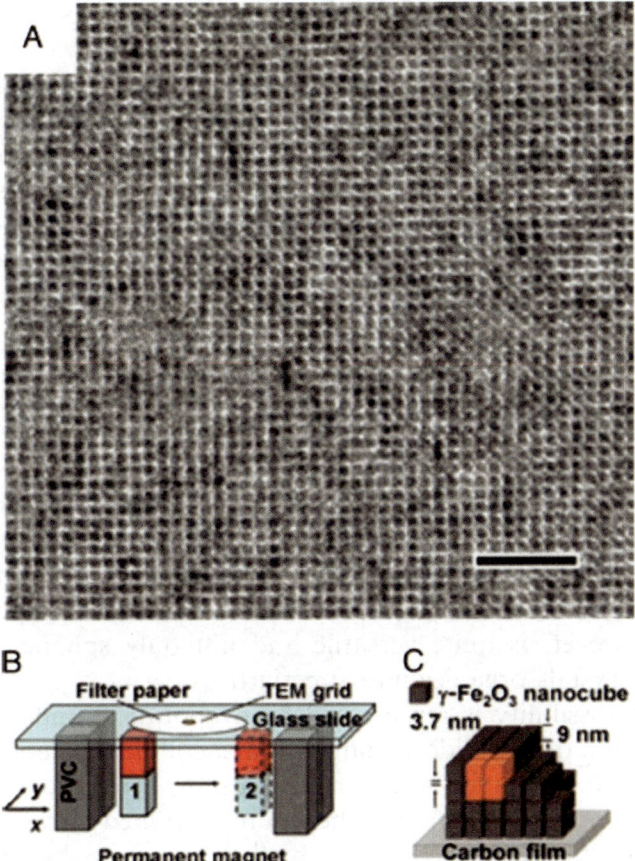

Figure 7.5 Magnetic field-induced self-assembly of oriented superlattices from maghemite nanocubes. (A) TEM image of magnetic-field-induced self-assembly of maghemite nanocubes into an oriented superlattice. (Scale bar: 100 nm.) (B) Schematic presentation of the magnetic-field-induced self-assembly procedure. A single permanent magnet was first moved slowly from point 1 to point 2 and then moved in the orthogonal direction immediately (within 2 min) after one or several drops of the nanocube dispersion was placed onto the TEM grid. (C) A schematic representation of a simple cubic superlattice formed by magnetic field-induced self-assembly. (Reprinted with permission from A. Ahniyaz *et al.*, *PNAS*, 2007, **104**, 17570.)

applying external fields but the colloidal droplet is the medium where self-organization occurs. It has been emphasized that droplet evaporation is not the only route to produce superlattices and that several methods have been developed. These methods try to fulfil the stringent requirements for producing ordered domains over a large area. Now we will focus our attention on some examples where evaporation induces organization without using a droplet.

An alternative method is to spread a colloidal droplet on a substrate using a controlled process; one possibility is the so-called 'doctor blade casting' method. A droplet is cast on a substrate and a moving scraping blade, with a constant speed, spreads the liquid on the substrate. The fast speed by which the blade moves favors the uniform spreading of the solution, which leaves behind a thin liquid film and this prevents any possible shrinkage of the wetted area such as in the case of an evaporating droplet.

This process has also been applied for spreading colloidal solutions to form superlattices.[21] Self-assembly *via* doctor blade deposition has been demonstrated for monodispersed magnetite heterostructures formed by a wüstite (FeO_x) core and a cobalt ferrite ($CoFe_2O_4$) shell. Also, in this case, the liquid−air interface is the place where self-assembly starts (**Figure 7.6**). Order arises during solvent drying and this depends on the nanoparticles' dimensions, concentration, type of solvent, amount of stabilizer, and also the substrate's properties. This process, however, is quite versatile and not only spherical, but also cubic nanocrystals organize into superlattices.

Another possibility is represented by evaporation under controlled conditions. If a quartz slide is immersed in a colloidal solution, the very slow evaporation under a controlled atmosphere and temperature allows for the formation of a hexagonal ordered array of packed nanoparticles[22] (**Figure 7.7**). This method is slow but allows for achieving quite large ordered areas and further allows for modulating the lattice constant by using alkanethiolate stabilizers of different alkyl chain lengths. Changing the distance between the metallic nanoparticles affects the near-field coupling and the collective surface plasmon resonance, which shifts accordingly. The resonance peak position

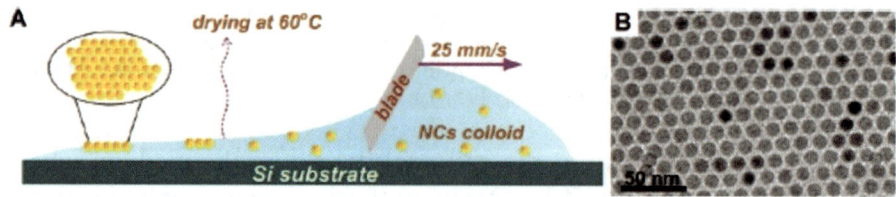

Figure 7.6   (A) Sketch of the doctor blade casting process where the nanocrystals assemble at the liquid−air interface. (B) TEM image of the nanocrystals with a size of 11 nm. (Adapted with permission from M. I. Bodnarchuk *et al.*, *ACS Nano*, 2010, **4**, 423. Copyright 2010, American Chemical Society.)

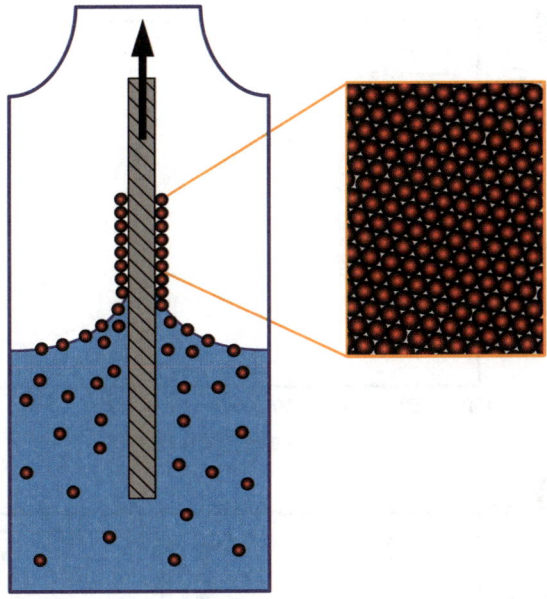

Figure 7.7   Formation of a superlattice *via* controlled evaporation on a substrate dipped in a colloidal solution.

depends on the interparticle gap and shifts to lower wavelengths with a decrease of the particle−particle distance (**Figure 7.8**).

7.6 BINARY NANOCRYSTALS

A superlattice that is formed of particles of different types is a material, or even better a 'metamaterial',[23,24] whose properties are not given by the combination of those of the individual particles but arise by the interaction of the different nanocrystals. A large variety of multi-component, and especially binary superlattices, have been obtained so far, which combines the properties of metal−metal, semiconductor−semiconductor, metallic−semiconductor, magnetic−metallic, and magnetic−magnetic nanoparticles. A wide range of structures of different stoichiometry together with a surprising long-range order has been obtained while the phases of these nanocrystal superlattices can be quite complex. If we look in more detail at the driving forces that are behind self-organization of binary superlattices things appear quite complex and a detailed analysis is beyond the scope of this book. However, we can say that a large variety of structures can be correlated to the balance of the different forces that can be part of the self-assembly process, such as van der Waals, dipole−dipole, entropy[25], ligand−ligand and Coulomb

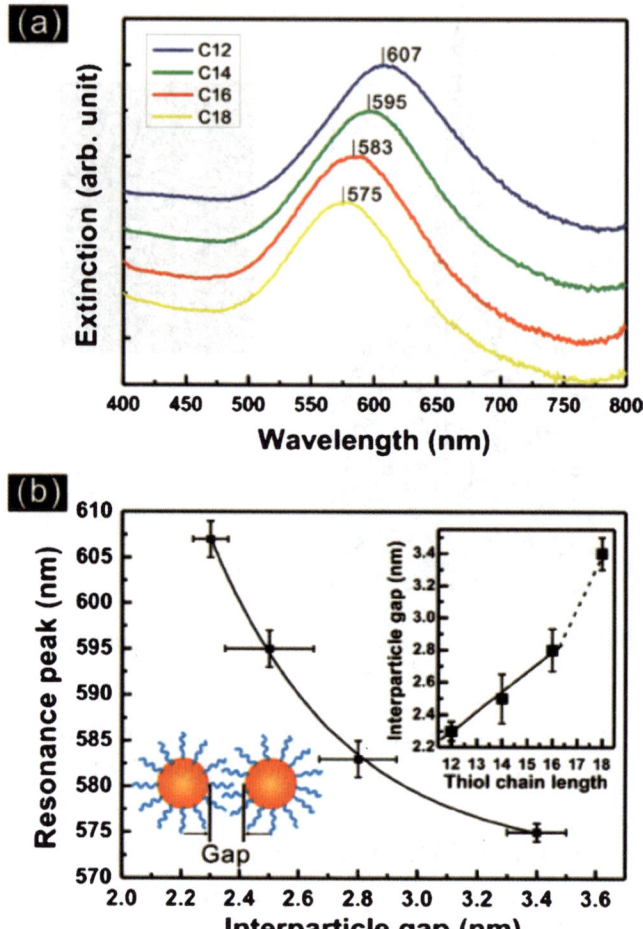

Figure 7.8 (a) Optical extinction spectra of four kinds of thiolate-stabilized nanoparticle superlattices with the same metal core size, but different in alkyl chain length. (b) Exponential dependence of the resonance peak position on the interparticle gap. The inset shows the relation between interparticle gap and thiolate chain length using the lattice constants obtained in Figure 7.2 and the known nanoparticle core size of 10.5 nm. (Reprinted with permission from C.-F. Chen *et al.*, *J. Am. Chem. Soc.*, 2008, **130**, 824. Copyright 2008, American Chemical Society.)

interactions. Of course the different contributions of these forces change with the composition of the particles and should be studied on a case-by-case basis while a general description of the process has not yet been obtained.

Binary superlattices can also be prepared by different methods. One of the most employed methods is the controlled evaporation of a colloidal solution on a substrate at a fixed angle.[26] The solid substrate, generally a TEM grid, is placed in a colloidal solution of binary nanoparticles that is put in a vial. The substrate is tilted to a fixed angle (30°) with respect to the surface of the drying dispersion. The evaporation rate of the solvent is adjusted by controlling the solvent vapor pressure during drying and the temperature (generally around 60–70 °C as a function of solvent employed). At the end of the evaporation process a binary superlattice forms on the TEM grid (**Figure 7.9**).

Several binary superlattices have been produced using this process. Some examples are the binary semiconductor PbSe (6 nm) and maghemite (γ-Fe$_2$O$_3$) superparamagnetic crystals (11 nm) that give AB$_{13}$-type (isomorphous with the intermetallic phase NaZn$_{13}$) superlattices[26] (**Figure 7.10**); the semiconductor PbSe and Pd metal superlattice[27]; the semiconductor PbSe (11 nm); and the semiconductor CdSe (5.8 nm).[28]

Again, it is important to stress that the non-equilibrium nature of self-assembly *via* evaporation implies that the process has some intrinsic instabilities, such as fluctuation in particle concentration or Marangoni flows, that are difficult to control. In the case of binary structures the interparticle interaction is also very important and the electrical charge of sterically-stabilized nanoparticles also has an important effect on the

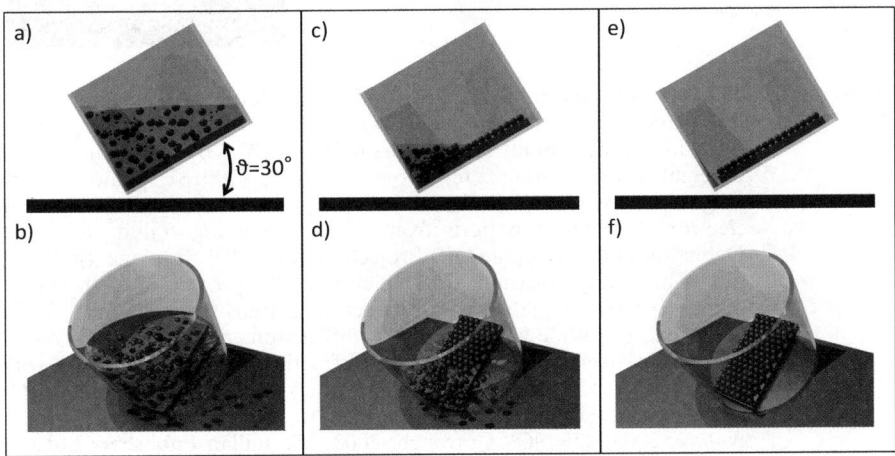

Figure 7.9 Schematic representation of the binary nanocrystal formation. Left: (a) and (b) TEM grid (copper) is placed in a mixed binary nanoparticle suspension in a vial with the substrate forming an angle of 30° with the surface of the drying dispersion. (c) and (d) Evaporation of the solvent from the suspension under reduced pressure at 70 °C and organization of nanoparticles. (e) and (f) The solvent is evaporated, the binary structures are formed on the TEM grid.

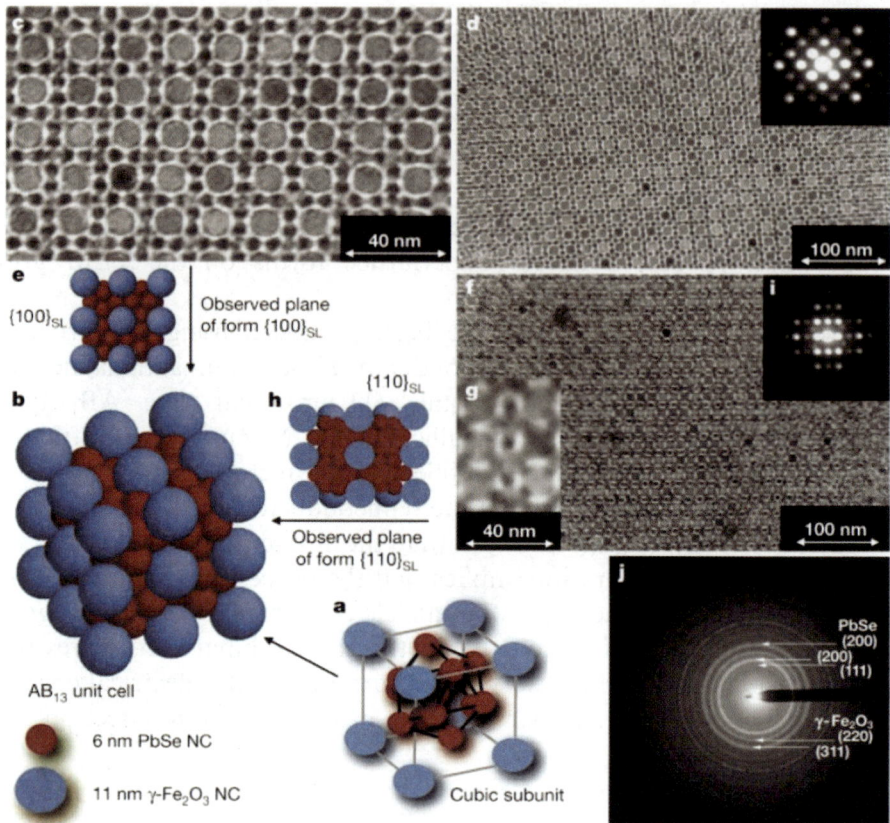

Figure 7.10   TEM micrographs and sketches of AB_{13} superlattices (isostructural with
intermetallic phase $NaZn_{13}$, SG 226) of 11 nm γ-Fe_2O_3 and 6 nm PbSe
NCs. (a) Cubic subunit of the AB_{13} unit cell. (b) AB_{13} unit cell built up
of eight cubic subunits. (c) Projection of a {100}SL plane at high
magnification. (d) As (c) but at low magnification; inset: small angle
electron diffraction pattern from a corresponding 6 mm^2 area. (e)
Depiction of a {100} plane. (f) Projection of a {110} SL plane. (g) As (f)
but at high magnification. (h) Depiction of the projection of the {110}
plane. (i) Small-angle electron diffraction pattern from a 6 mm^2 {110}
SL area. (j) Wide-angle electron diffraction pattern of an AB_{13}-
superlattice (selected-area electron diffraction of a 6 mm^2 area) with
indexing of the main diffraction rings for PbSe and γ-Fe_2O_3
(maghemite). (Reprinted with permission from: F. X. Redl *et al.*,
Nature, 2003, **423**, 968. Copyright 2003, Macmillan Publishers Ltd.)

binary superlattice stoichiometry. This is not the only contribution to
self-assembly because the stabilization of the superlattice is achieved
with the additional contributions of entropy, van der Waals interactions
and steric and dipolar forces. **Figure 7.11** shows 12 of the binary

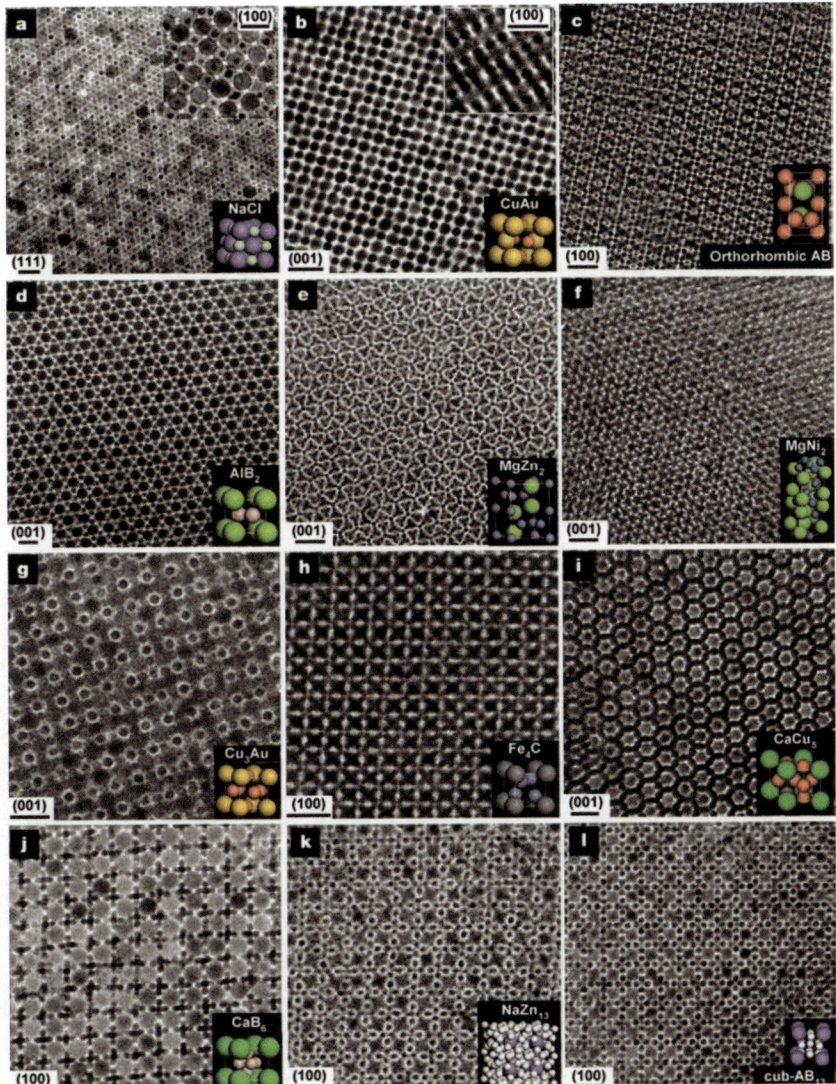

Figure 7.11 TEM images of the characteristic projections of the binary superlattices, self-assembled from different nanoparticles and modelled unit cells of the corresponding 3D structures. The superlattices are assembled from (a) 13.4 nm γ-Fe_2O_3 and 5.0 nm Au; (b) 7.6 nm PbSe and 5.0 nm Au; (c) 6.2 nm PbSe and 3.0 nm Pd; (d) 6.7 nm PbS and 3.0 nm Pd; (e) 6.2 nm PbSe and 3.0 nm Pd; (f) 5.8 nm PbSe and 3.0 nm Pd; (g) 7.2 nm PbSe and 4.2 nm Ag; (h) 6.2 nm PbSe and 3.0 nm Pd; (i) 7.2 nm PbSe and 5.0 nm Au; (j) 5.8 nm PbSe and 3.0 nm Pd; (k) 7.2 nm PbSe and 4.2 nm Ag; and (l) 6.2 nm PbSe and 3.0 nm Pd nanoparticles. Scale bars: (a)–(c), (e), (f), (i)–(l), 20 nm; (d), (g), (h), 10 nm. The lattice projection is labelled in each panel above the scale bar. (Reprinted with permission from: E. V. Schevchenko *et al.*, *Nature*, 2006, **439**, 55. Copyright 2006, Macmillan Publishers Ltd.)

superlattices obtained by self-assembly *via* evaporation of different combinations of nanoparticles.[29]

7.7 MORE THAN CRYSTALS...QUASICRYSTALS

Colloidal nanocrystals have shown an incredible capability to order into packed layered structures that can be defined as 'crystals of crystals' and that exhibit new properties, which differ greatly from those of single particles such as in binary superlattices. The surprises, however, do not stop here as self-assembly can challenge our common understanding of order. Moving beyond the order of crystalline materials with a regular arrangement of atoms, there are other solids that have long-range order without the 3D periodicity and translational symmetry that characterizes conventional crystalline materials. The atomic organization of these solids allows 'forbidden' rotational symmetries

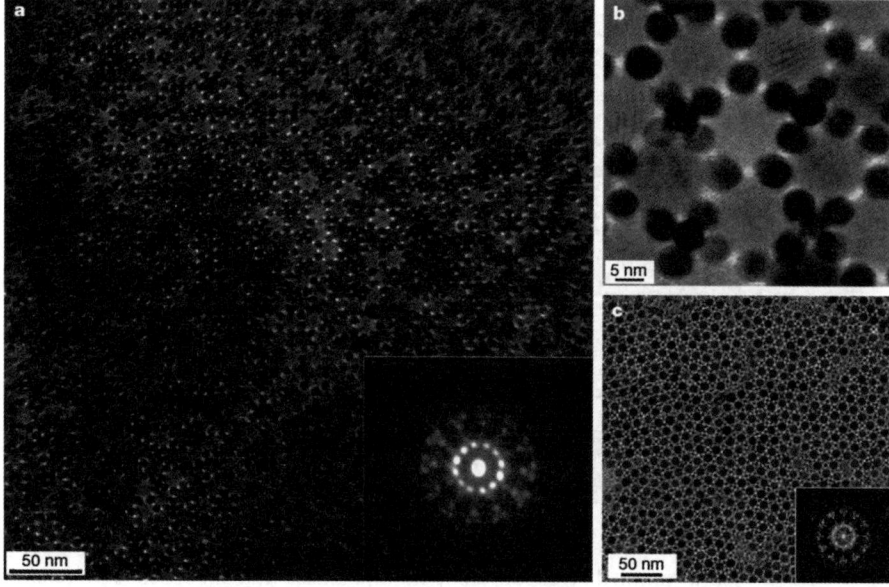

Figure 7.12   Dodecagonal quasicrystals self-assembled from spherical nanoparticles. (a) TEM image of a quasicrystalline superlattice self-assembled from 13.4 nm Fe_2O_3 and 5 nm Au nanocrystals. Inset, selected-area electron diffraction pattern with non-crystallographic 12-fold rotational symmetry measured from a 6 mm^2 domain. (b) Magnified view of a dodecagonal nanoparticle quasicrystal. (c) Dodecagonal quasicrystalline superlattice self-assembled from 9 nm PbS and 3 nm Pd nanocrystals. Inset, fast-Fourier transform pattern of the quasicrystalline superlattice. (Reprinted with permission from: D. V. Talapin *et al.*, *Nature*, 2009, **461**, 964. Copyright 2009, Macmillan Publishers Ltd.)

such as 5-fold, 8-fold, 10-fold and 12-fold rotations and, at the same time, still exhibits sharp diffraction peaks. These materials have been defined as quasicrystals[30] and have definitely been entered into the family of ordered structures. The quasicrystal universe has been widened by the observation that even soft matter such as liquid crystals, surfactants and polymers may show quasi-periodicity with forbidden symmetries.[31]

By using the same method as described previously for the fabrication of binary superlattices, quasicrystalline order has been observed in self-assembled nanoparticle superlattices. It has been observed that by slow evaporation of binary colloidal solutions of different compositions, binary aperiodic superlattices of nanocrystals are obtained[32] (**Figure 7.12**). The possibility of achieving quasicrystalline organization by different types of particles suggests that self-assembly is not given by a unique combination of interparticle interactions. It appears as a more general sphere-packing phenomenon where entropy and particle−particle potentials drive the organization.

REFERENCES

1. Y. Yin and A. P. Alivisatos, *Nature*, 2005, **437**, 664.
2. D. V. Talapin, J.-S. Lee, M. V. Kovalenko and E. V. Shevchenko, *Chem. Rev.*, 2010, **110**, 389.
3. M. A. El-Sayed, *Acc. Chem. Res.*, 2004, **37**, 326.
4. N. Zheng, J. Fan and G. D. Stucky, *J. Am. Chem. Soc.*, 2006, **128**, 6550.
5. F. Li, D. P. Josephson and A. Stein, *Angew. Chem., Int. Ed.*, 2011, **50**, 360.
6. P. Podsiadlo, B. Lee, V. B. Prakapenka, G. V. Krylova, R. D. Schaller, A. Demortiere and E. V. Shevchenko, *Nano Lett.*, 2011, **11**, 579.
7. J. B. He, P. Kanjanaboos, N. L. Frazer, A. Weis, X. M. Lin and H. M. Jaeger, *Small*, 2010, **6**, 1449.
8. Z. L. Wang, S. A. Harfenist, R. L. Whetten, J. Bentley and N. D. Bundling, *J. Phys. Chem. B*, 1998, **102**, 3068.
9. X. M. Lin, H. M. Jaeger, C. M. Sorensen and K. J. Klabunde, *J. Phys. Chem. B*, 2001, **105**, 3353.
10. S. Narayanan, J. Wang and X.-M. Lin, *Phys. Rev. Lett.*, 2004, **93**, 135503-1.
11. T. P. Bigioni, X.-M. Lin, T. T. Nguyen, E. I. Corwin, T. A. Witten and H. M. Jaeger, *Nat. Mater.*, 2006, **5**, 265.
12. T. Nishio, K. Niikura, Y. Matsuo and K. Ijiro, *Chem. Commun*, 2010, **46**, 8977.
13. M. Achermann, M. A. Petruska, S. A. Crooker and V. I. Klimov, *J. Phys. Chem. B*, 2003, **107**, 13782.
14. M.-H. Lin, H.-Y. Chen and S. Gwo, *J. Am. Chem. Soc.*, 2010, **132**, 11259.
15. S. I. Lim and C.-J. Zhong, *Acc. Chem. Res.*, 2009, **42**, 798.

16. V. Santhanam, J. Liu, R. Agarwal and R. P. Andres, *Langmuir*, 2003, **19**, 7881.
17. M. N. Martin and S.-K. Eah, *Mater. Res. Soc. Symp. Proc.*, 2009, **1113**, F03–01.
18. S.-K. Eah, *J. Mater. Chem.*, 2011, **21**, 16866.
19. M. N. Martin, J. I. Basham, P. Chando and S.-K. Eah, *Langmuir*, 2010, **26**, 7410.
20. A. Ahniyaz, Y. Sakamoto and L. Bergstrom, *Proc. Nat. Acad. Sci.*, 2007, **104**, 17570.
21. M. I. Bodnarchuk, M. V. Kovalenko, S. Pichler, G. Fritz-Popovski, G. Hesser and W. Heiss, *ACS Nano*, 2010, **4**, 423.
22. C.-F. Chen, S.-D. Tzeng, H.-Y. Chen, K.-J. Lin and S. Gwo, *J. Am. Chem. Soc.*, 2008, **130**, 824.
23. A 'metamaterial' has properties that cannot be found in nature, such as a negative refractive index.
24. N. Engheta and R. W. Ziolkowski, *Metamaterials: Physics and Engineering Explorations*, John Wiley and Sons, USA, 2006.
25. M. D. Eldridge, P. A. Madden and D. Frenkel, *Nature*, 1993, **365**, 35.
26. F. Redl, K.-S. Cho, C. B. Murray and S. O'Brien, *Nature*, 2003, **423**, 968.
27. E. V. Shevchenko, D. V. Talapin, S. O'Brien and C. B. Murray, *J. Am. Chem. Soc.*, 2005, **127**, 8741.
28. W. H. Evers, B. De Nijs, L. Filion, S. Castillo, M. Dijkstra and D. Vanmaekelbergh, *Nano Lett.*, 2010, **10**, 4235.
29. E. V. Shevchenko, D. V. Talapin, N. A. Kotov, S. O'Brien and C. B. Murray, *Nature*, 2006, **439**, 55.
30. D. Levine and P. J. Steinhardt, *Phys. Rev. Lett.*, 1984, **53**, 2477.
31. T. Dotera, *J. Polym. Sci. B: Polym. Phys.*, 2012, **50**, 155.
32. D. V. Talapin, E. V. Shevchenko, M. I. Bodnarchuk, X. Ye, J. Chen and C. B. Murray, *Nature*, 2009, **461**, 964.

CHAPTER 8

Shaping and Ordering the Porosity Through Self-assembly

In previous chapters we have seen plenty of examples of self-assembly that are driven by evaporation phenomena. These processes have something in common; the order is achieved by the organization of small objects of predefined shape and dimension on the nanoscale. In most cases, controlling the process is the key for achieving self-assembly.

We could however change our perspective a little. What if self-assembly is again driven by evaporation but we start from single molecules? The point of arrival is the same, organized nano-objects, but in this case they are not preformed. The difficulty of managing the overall process appears to be much higher. In this chapter, therefore, we introduce more complexity into the system that should be able to self-assemble, using only chemical−physical processes, into a structure with ordered porosity. We have selected for this chapter a specific type of mesoporous material, mesoporous ordered films, because the fast evaporation conditions during deposition make them a much more interesting case for a better understanding of self-assembly.

8.1 WHAT TYPE OF POROSITY? ESCAPING FROM A ZEOLITE TRAP

In Chapter **5**, we became acquainted with a material whose properties rely on its porous ordered structure: photonic crystals. In this case the length scale of interest is microns and the self-assembly of templating microparticles through controlled evaporation processes is also a possible fabrication route. At a lower scale, we find other porous

Water Droplets to Nanotechnology: A Journey Through Self-Assembly
By Plinio Innocenzi, Luca Malfatti and Paolo Falcaro
© P. Innocenzi, L. Malfatti and P. Falcaro 2013
Published by the Royal Society of Chemistry, www.rsc.org

materials, which also exhibit ordered porosity, the zeolites, with pores in the 0.3–1.5 nm range. At this point it is important to define porous materials on the basis of pore size. At the lower limit we have micropores (< 2 nm), then mesopores (between 2 and 50 nm) and finally macropores (> 50 nm).[1] Zeolites are inorganic materials whose microporosity has a well-defined shape and dimension and represent the most important example of microporous materials.[2] They are of paramount importance for catalysis, especially in the petrochemical industry, but their porosity range limits the capability of absorbing molecules of larger dimensions. In an attempt to 'escape from the zeolite trap' with the limit of micropores, researchers from Mobil in 1992 finally obtained mesoporous inorganic materials using surfactant molecules as templates.[3,4] This material is known as MCM (Mobil Crystalline Materials) and has filled the gap between zeolites and photonic crystals (**Figure 8.1**).

8.2 WE NEED A TEMPLATE

The basic idea is to produce, *via* self-assembly, a material with ordered pores in the mesoporous range and, just to make matters more complex, nothing should be preformed, neither the template or the building blocks of the material. This means that it is necessary to start from precursors that are able to form during an evaporative process: a template of well-defined shape and dimension. The template must not be affected by the synthesis, it must self-organize into an ordered array, it should form a suitable interface for the formation of the inorganic network and finally it should be easy to remove at the end of the process (**Figure 8.2**). If we look at all these properties it seems quite a hard task

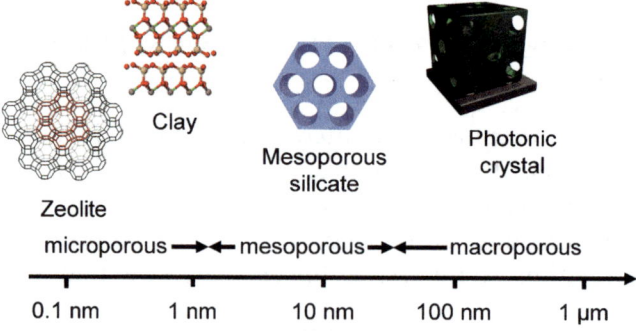

Figure 8.1 The length scale of porous materials, from micropores to macropores (Adapted with permission from S. Förster *et al.*, *Angew. Chem., Int. Ed.*, 2002, **41**, 688. Copyright 2002, John Wiley & Sons.)

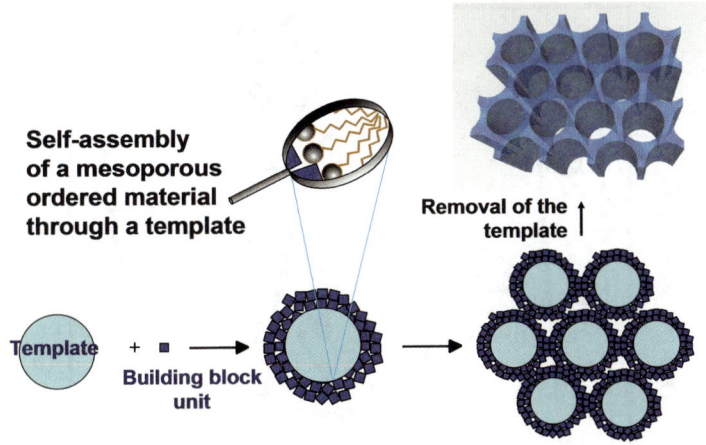

Figure 8.2 A scheme of formation of mesoporous materials *via* a template.

but fortunately these requirements are easily satisfied by quite well known molecules such as surfactants and also block-copolymers. These species have to be amphiphilic and, therefore, under the right conditions of concentration and temperature, they can form different types of micelles, which have been revealed to be quite stable and in turn tend to self-assemble during evaporation. Surfactants have shown this type of property: they have a lyophilic (solvent-loving) head group and a lyophobic (solvent-hating) tail and in water they form micelles with hydrophilic head groups on the outer surface and hydrophobic tails that point towards the center. Some common surfactants are ionically charged, such as cetyltrimethylammonium bromide $(CH_3(CH_2)_{15} N(CH_3)_3^+Br^-$ (CTAB), and the relative stability of their micelles is given by the balance between the repulsion of the charged head groups and minimization of the interactions of the hydrophobic tails with water. The surfactants not only form a supramolecular stable structure but also allow, within some extents, to design the shape of the micelles by changing the synthesis conditions. As it is shown in the schematic phase diagram of CTAB in water, with the increase of the surfactant concentration, c, the shape of the micelle changes and different structures form (**Figure 8.3**).[5] Below a specific concentration level, which represents the critical micelle concentration (CMC), the surfactant molecules are free to move and remain in solution as single molecules. When c is larger than CMC (CMC 1 in **Figure 8.3**), small spherical micelles form. At higher concentrations (CMC 2), rod-like elongated cylindrical micelles grow because the solvent available between the micelles decreases and it is easier for spherical micelles to

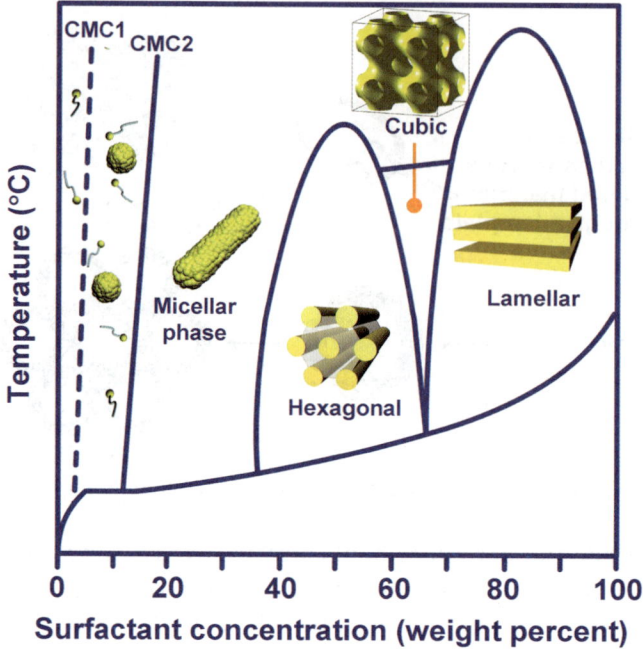

Figure 8.3 Phase diagram of a surfactant, CTAB, in water as a function of concentration and temperature. (Adapted with permission from N. K. Raman *et al.*, *Chem. Mater.*, 1996, **8**, 1682. Copyright 1996, American Chemical Society.)

come in contact to coalesce. By increasing the concentration even more, the rod-like micelles form hexagonal close-packed liquid crystalline (LC) or cubic arrays as well as lamellar phases. If the concentration is very high, in some cases, the surfactants completely reverse the molecular arrangement of the micelle and form inverse phases with water, which is solubilized at the core part and the head groups in turn point inward. Different surfactants will form different micelles as a function of the concentration but is predicting the final self-assembled structure somehow possible? The geometry of the molecules helps to make reasonable predictions and a simple model of micellar structures, which is based on geometrical considerations, has been developed.[6] This model considers the space occupied by the head and the tail of the surfactant and introduces the critical packing parameter (CPP), which is defined as:

$$CPP = \frac{V_H}{a_0 l_c} \qquad (8.1)$$

where V_H is the volume of the hydrophobic portion of the molecule, a_o, the effective optimal head group surface area, and l_c the critical length of the hydrophobic tail. The value of l_c depends on the extension of the chain and can be calculated by the number of carbon atoms, n, in the tail of the surfactant by $l_c \leqslant 1.5 + 1.265n$ Å. The value of CPP gives an indication of the type of micelle that should be expected, the higher the value of CPP, the lower the curvature of the micelle (**Table 8.1**).

There are plenty of types of surfactants that can be used as templates for mesoporous materials, they differ in size, shape and charge and are classified into three main groups: anionic, with the hydrophilic group carrying a negative charge; cationic, where the hydrophilic group has instead a positive charge; and finally, non-ionic, without any charge.

There is another category of molecules beside surfactants that show an impressive capability of self-assembly and can be therefore used for the synthesis of mesoporous materials, the block copolymers.[7,8] They also easily self-assemble into micelles of well-defined shapes and sizes.[9] The molecules are formed by chemically-connected blocks with different properties, for instance one block can be hydrophilic and the other hydrophobic; this creates the conditions for self-organization of the single blocks into nanoscale domains. When the chain lengths and the volume fraction reach a certain critical value the block copolymers form micelles of different shapes, such as the case of the surfactants that we have just seen. The size of the micelle depends on the length of the blocks; a systematic correlation between the aggregation number, Z, which corresponds to the number of block copolymers in a micelle, and the degree of polymerization of the insoluble block, N_A, exists:

$$Z = Z_0 \frac{N_A^\alpha}{N_B^\beta} \qquad (8.2)$$

where $\alpha = 2$ and $\beta = 0.8$, while Z_0 depends on the enthalpy of the

Table 8.1 Expected micellar structure as a function of critical packing parameter (CPP).

CPP	Type of structure
< 0.33	Spherical micelles
0.33–0.5	Cylindrical or rod-like shaped micelles
0.5–1.0	Bilayer or vesicles
1–2	Bilayer or membranes
2–3	Inverse cylindrical micelles
> 3	Inverse spherical micelles

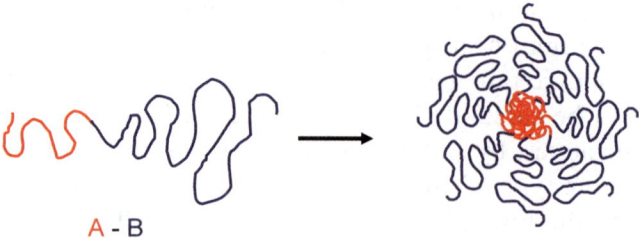

A - B

Figure 8.4   Formation of a micelle from an A-B type diblock copolymer with aggregation number $Z = 8$. (Adapted with permission from S. Förster *et al.*, *Angew. Chem., Int. Ed.*, 2002, **41**, 688. Copyright 2002, John Wiley & Sons.)

mixing between the insoluble polymer block A and the solvent. In **Figure 8.4** we can see the formation of a micelle with aggregation number $Z = 8$ from a copolymer of two blocks (A and B type). The law in eqn (8.2) works surprisingly well for describing the formation of micelles from diblock, triblock, graft- and star-block copolymers and even for surfactants. This is also an indication that the self-assembly mechanism behind the different molecules is basically the same.

The most commonly used block copolymers for the self-assembly of mesoporous materials are generally diblock (commercial name Brij) or triblock copolymers (commercial name Pluronic) and are composed of hydrophilic (polyethylene oxide, PEO) and hydrophobic (polypropylene oxide, PPO) blocks (**Figure 8.5**).[10]

With respect to surfactants, block copolymers in general allow for obtaining micelles of larger dimensions and are easier to remove. Of course the choice of the template is not without consequence and the properties of the final mesoporous materials directly depend on this.

There is a final observation about the templates that we have just described that is worth mentioning and it is that they are all 'soft'.[11] This means that they are compliant enough to allow organization and rearrangements at least until they are not finally removed. This is an important property that, as we will see, also plays an important role during self-assembly.

8.3   WE NEED BRICKS

Micelles, as we have just seen, represent an impressive template with the capability of forming nano-objects of a specific shape which also self-assemble into organized arrays. They are, however, soft organic objects, and even if they are able to form ordered arrays, these have limited

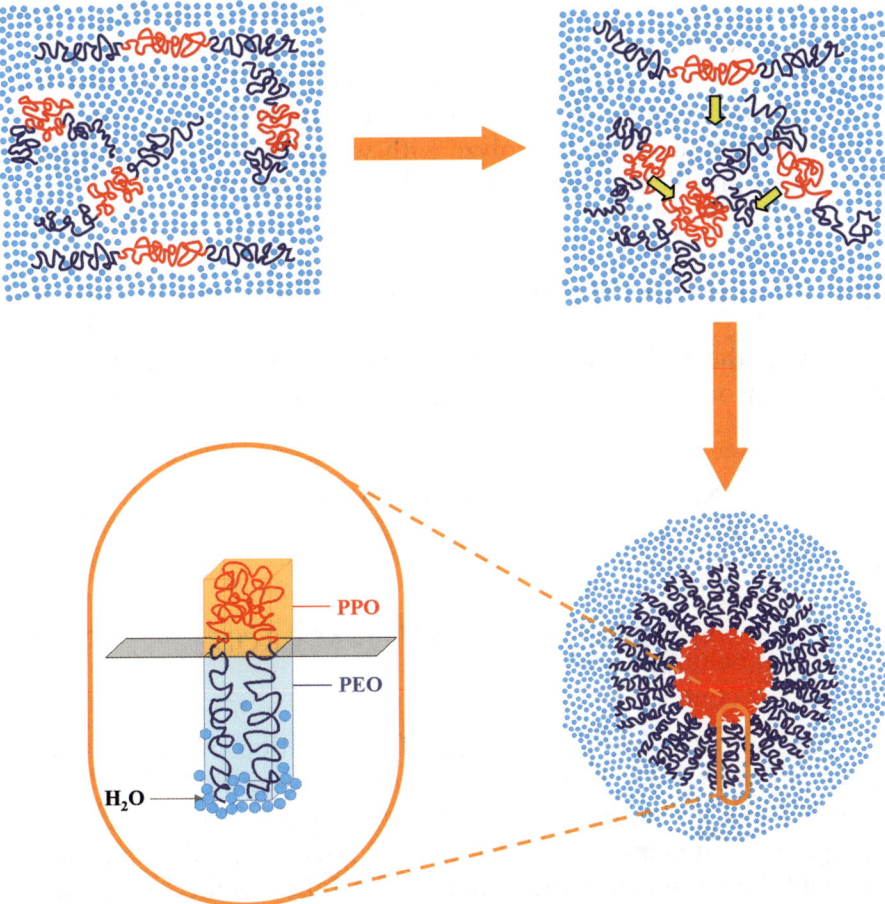

Figure 8.5   Formation of a micelle from a block copolymer of the A−B−A type. The copolymer is formed by alternating a hydrophilic polyethylene oxide (PEO) block, a hydrophobic block of polypropylene oxide (PPO) and another hydrophilic block of PEO. In water these block copolymers form micelles with a hydrophobic core (PPO) and an outer hydrophilic region (PEO).

chemical and mechanical stability. It is now necessary to think of the second step (making a mesoporous material). We still need the bricks, or such as in the present case, nano-building blocks. They cannot be preformed objects, however, as they should also form and connect during the evaporation process. The best way of getting these building units is by using chemistry. The synthesis, however, needs to be carried out in the solution where we have solubilized the precursor for the templates. Quite fortunately such a possibility exists and is offered by sol−gel chemistry; a low temperature route to prepare oxides and hybrid materials.[12] To achieve sol–gel chemistry we need an appropriate precursor, solvent, water for the hydrolysis reaction and a catalyst. All these components can be part of the same precursor solution that contains the surfactants.[13] The most common precursors are metal alkoxides that hydrolyze with water forming metal hydroxy species and producing alcohol molecules as by product:

$$M-OR + H_2O \rightarrow M-OH + ROH \qquad (8.3)$$

where M is the metal and R is the organic group. The hydroxy species can react through condensation reactions and form oligomers *via*:

$$M-OH + HO-M \rightarrow M-O-M + H_2O \qquad (8.4)$$

or:

$$M-OR + HO-M \rightarrow M-O-M + ROH \qquad (8.5)$$

These 'simple' reactions allow for the formation of oxo-clusters of fractal shape that finally condense with the evaporation of the solvent to pass from the state of sol (a suspension containing particles or aggregates of dimension smaller than 1 micron) to the state of gel where all the clusters are interconnected and there is a coexistence of a liquid and solid phase. This chemistry also has the advantage of being extremely flexible and allows for tailoring within some limits the size, shape, dimension and composition (mixed oxides are obtained *via* controlled co-condensation of different metal alkoxides) of the oxide clusters that gives some room for designing the self-assembly process.

Another advantage of sol–gel chemistry is that we are not limited to oxides for making mesoporous materials. In fact in the case of silicon alkoxides the hydrolytic stability of the chemical bond between silicon and carbon, Si–C, allows for producing a new family of materials that are organic–inorganic hybrids with a large variety of organic functional molecules. This means that we can obtain *via* self-organization purely inorganic mesoporous materials of different compositions and organic–inorganic hybrids.[14,15]

8.4 THE RACE TO ORDER

Now that we have the template and the bricks we can finally allow self-assembly to occur and our work should be almost finished. However, our task is not completely over. Before starting we should be sure that the conditions for self-assembly have really been achieved. When the system containing the solvent, the amphiphilic molecules, and the sol−gel precursor starts evaporating, two different chemical−physical processes begin to compete with each other. One is the formation of the micelles and the other is the condensation of the oxo-clusters. If the kinetics of these processes does not follow the proper order the building units will self-condense giving a separate structure from the micelles.[16] The condensation rate should be adjusted, therefore, to follow the formation of the micelles and the rise of a hybrid interface. In this type of race-to-order it is necessary that all the participants maintain a predefined speed and follow very carefully the instructions. At the end of the race, an ordered structure is obtained only if the kinetic constants of the process follow this hierarchy:[17]

$$k_{\text{inter}} > k_{\text{org}} > k_{\text{inorg}} \qquad (8.6)$$

where k_{inter}, k_{org} and k_{inorg} are the relative rates for interface formation, organic array assembly and inorganic polycondensation, respectively. In general, the main problem that needs to be solved is to slow down the condensation of the oxide clusters. In the case of silica the condensation reactions are slower under highly acidic conditions (pH < 2) and this is enough to allow self-assembly. This also works quite well for most of the oxides.

We can now follow, in more detail, what happens in the evaporation stage of a liquid film, which contains all the ingredients for self-assembly and how this process drives the organization. We need to stress again that during the deposition of the film the evaporation is fast and this makes the conditions for organization quite critical. A common technique for the deposition of a film from a liquid phase is dip-coating. A substrate is immersed in the solution and then withdrawn at a controlled rate. It is this speed that affects the thickness of the final film. A scheme of the process is shown in **Figure 8.6(a)**. The solution typically contains the sol−gel oligomers, ethanol and water as solvents, and the surfactant or the block copolymer at a concentration which is lower than CMC.[18] When the substrate is pulled up the evaporation begins and at a certain point the template forms micelles. At higher

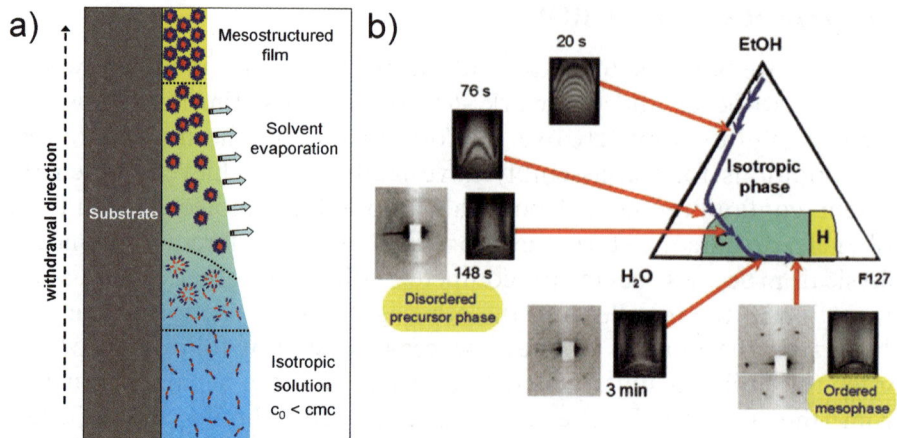

Figure 8.6   (a) Formation of a thin mesostructured film upon dip-coating. (b)
Simplified system trajectory of the formation of a mesostructured TiO$_2$
film plotted along a ternary diagram non-ionic surfactant (F127)/
ethanol/water. Typical interferometric and SAXS patterns are shown for
each point. (Reprinted with permission from G. J. A. A. Soler-Illia *et al.*,
Chem. – Eur. J., 2006, **12**, 4478. Copyright 2006, John Wiley & Sons.)

concentrations a liquid crystalline phase is formed and at the same time
the building units pile up on the surface and condense to form an
interconnected network around the ordered templating structure of the
micelles.

To follow in more detail what is happening during film formation a
system trajectory, obtained by *in situ* small-angle X-ray scattering
(SAXS) measurements, can be used (**Figure 8.6(b)**).[19–22] The scheme
shows a ternary diagram with ethanol, water and a triblock copolymer
(Pluronic F127). *In situ* snapshots taken with interferometry shed light
and SAXS patterns[23] follow the evolution of the films during
evaporation. At the beginning of the process ethanol evaporates from
the film, which becomes enriched in water, and for several seconds this
is the main experimental fact and the system does not organize. After
this elapsed time, however, the SAXS analysis reveals the formation of
a circular pattern, which indicates a disordered structure, which is
sometimes called 'spaghetti-like' where order is not yet reached but
some correlation at a smaller scale is present. In the final step this
disordered precursor structure finally evolves into a well-defined phase,
with 2D or 3D organization. The last stage is the removal of the
template, which can be achieved by several easy routes, such as thermal
treatment, UV exposure or chemical extraction. At the end of the

process an oxide film with porosity that replicates the same order achieved by the template is obtained.

This is a very simplified description but works well for giving a general comprehension of the phenomenon that is by far more complex than the self-assembly of predefined objects that we have seen in previous chapters.

8.5   WHAT TYPE OF ORDER?

After we know how to self-assemble mesoporous films, the next important questions are what type of ordered structures can we obtain and how can we design them? The difficult point is that the system is not an equilibrium and that, during solvent evaporation, sol−gel reactions proceed quickly and change the conditions, such as viscosity, and the critical micelle concentration, which are also reached at a certain time in the process and cannot be described as *a priori*.[24] This is the reason why making mesoporous films is very much trial and error, but even if starting from an empirical basis a very large variety of ordered phases has been obtained and a summary of the most common mesophases that have been reported so far is shown in **Figure 8.7**.[25] The simplest case is the worm-like disordered phase, which has already been described in the previous section. It has only a local order and is the result of a poor capability of the system to self-assemble. In silica it forms when the condensation of the oxide network is too extended or the amount of surfactant too low. True order is observed when the micelles are able to template mesophases with 2D and 3D organization. The most common 2D structures are lamellar or 2D hexagonal (2D Hex). The lamellar phase is, in general, not stable and of small interest while 2D Hex, such as the *p6mm* and *c2mm* (this is the notation for classification of crystals in the space group), because of the possibility of orienting the channels, are quite appealing for several applications. A higher symmetry is shown by 3D structures that can be hexagonal such as the 3D Hex phase, *P6₃/mmc*; or the primitive cubic phase, *Pm$\bar{3}$n*; the bicontinuous cubic, *Pn$\bar{3}$m*; the body centred cubic phase, *Im$\bar{3}$m*; the double-gyroid cubic phase, *Ia$\bar{3}$d*; the face-centred cubic phase, *Fm$\bar{3}$m;* or the tetragonal-like body-centred tetragonal phase, *I4/mmm* or rhombohedral, *R$\bar{3}$m*.

The overall organization process is kinetically controlled and the final order depends on the processing parameters, such as the withdrawal rate of the substrate and the relative humidity within the deposition room. In any case, the nature of the interface, such as micelle curvature

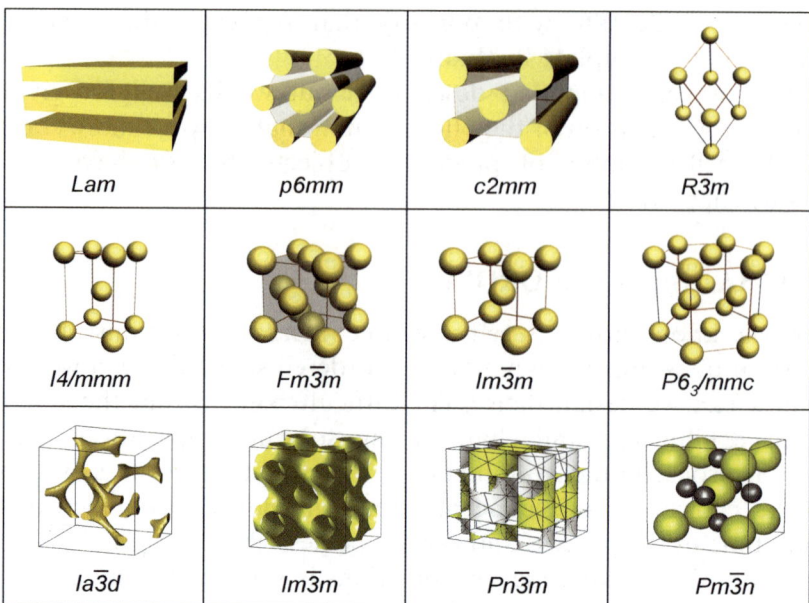

Figure 8.7 Pore structures obtained *via* self-assembly templating in mesoporous films: 2D (Lam = lamellar, *p6mm*, and *c2mm*), 3D ($R\bar{3}m$, *I4/mmm*, $Fm\bar{3}m$, $Im\bar{3}m$, *P6₃/mmc*, and $Pm\bar{3}n$) and bicontinuous ($Ia\bar{3}d$, $Im\bar{3}m$ and $Pn\bar{3}m$). (Reprinted with permission from P. Innocenzi *et al.*, *Chem. Mater.*, 2009, **21**, 2555. Copyright 2009, American Chemical Society.)

and the interaction between the building units and the micelle has a predominant role.

8.6 'CRYSTALS' OF PORES

Let us now challenge the concept of order and crystallinity we are used to. Can we indeed define the ordered array of pores in a mesostructured film as a 'crystalline-like' structure? An answer is given by X-ray diffraction analysis: even if an oxide mesoporous film is not crystalline it gives diffraction patterns at low angles when the pores are ordered. This is due to the presence of periodic voids or, if the material is not calcined, to the periodic contrast in the electronic density in correspondence of the surfactant periodic array that generates the X-ray diffraction at low angles. The dimensions of the unit cells are, however, larger in comparison with ordinary crystalline materials. Instead of a lattice of atoms (0.1–0.2 nm), we have an ordered structure of voids and walls (both in the 2–6 nm range), and therefore the lattice parameters will be between 4 and 12 nm, a much higher range with respect to the typical dimensions of a crystalline structure ($\sim$ 0.5 nm). However, to make it more complex, in oxides the pore walls can also crystallize after thermal calcination[26] and if the annealing process does not disrupt the ordered porous structure, low-angle and high-angle diffraction patterns will both be observed. The mesoporous films, therefore, can be treated as a kind of porous crystal, but by how much is this order extended? In conventional crystalline materials the structure is generally composed of many small crystal of different orientations, which join to each other through grain boundaries. In controlled, and special, cases a monocrystal is observed. For mesoporous materials the analogy is surprising and different types of analysis have revealed that there is a distribution of ordered porous domains, 'polycrystalline-like', which are well oriented along the substrate in the case of 2D Hex phases, but all possible in-plane orientations are also possible (**Figure 8.8**). The case of 3D porous structures is similar and the formation of ordered domains of some microns is generally observed.

In most of the cases ordered porous domains with random orientations, which generally are extended for some microns, form mesoporous films. However the system can be even more complicated because in the same film at the end of the process different mesophases can coexist, such as disordered domains or lamellar structures together with cubic or hexagonal mesophases. This case is clearly favored by the interfaces (solid−liquid and liquid−air) as shown by the high resolution TEM image (**Figure 8.9**) of a mesoporous silica film section. Different types of ordered domains, hexagonal (at the substrate–film interface), worm-like in the middle, and cubic at the film−air interface coexist in the same sample.[27]

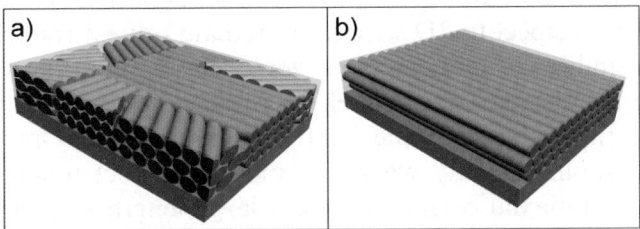

Figure 8.8 2D Hex ordered porous film with 'polycrystalline-like' domains (a) and 'monocrystal-like' structure of aligned pores with in plane orientation (b). (Reprinted with permission from P. Innocenzi *et al.*, *Chem. Mater.*, 2009, **21**, 2555. Copyright 2009, American Chemical Society.)

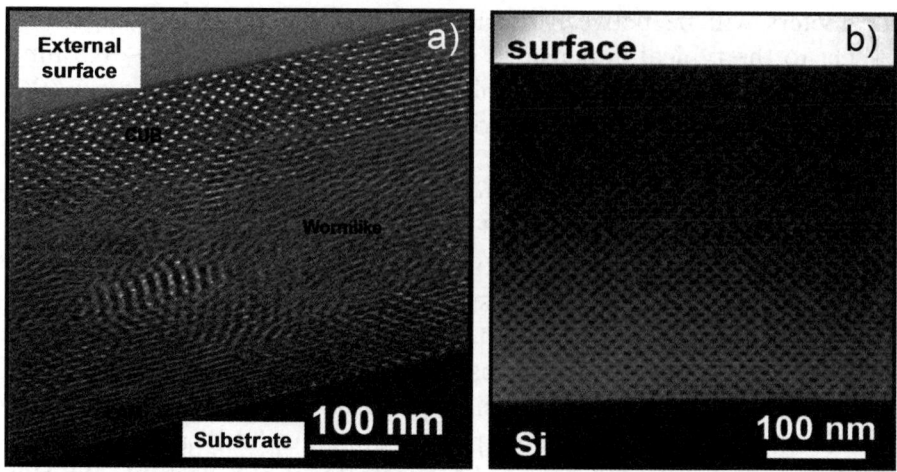

Figure 8.9   (a) HRTEM image of a mesoporous silica film showing the coexistence of hexagonal (Hex), wormlike, and cubic (Cub) pore domains. (b) HRTEM image of a hybrid mesoporous silica film with highly ordered tetragonal pore structure. (Reprinted with permission from P. Falcaro *et al.*, *J. Am. Chem. Soc.*, 2005, **127**, 3838. Copyright 2005, American Chemical Society.)

8.7   DESIGNING ORDER: ORIENTING THE PORES

The self-assembly process for obtaining mesostructured films, beside its complexity also presents some flexibility that allows the design of the pore structure. The final challenge to face, therefore, is trying to orient the pores. There are some 3D structures with a high degree of symmetry, such as the gyroids, which are particularly interesting for applications where full accessibility of the pores and mobility within them is required.[28] They have a high degree of symmetry and beside some difficulties in their synthesis they do not need to be oriented. However, mesoporous films with a 2D phase organization, such as the 2D rectangular *c2mm* and the 2D Hex *p6mm*, even if they have a lower symmetry with respect to 3D structures, remain quite attractive because of the possibility of aligning the mesophase along a preferential direction. Several types of channel orientation could be of interest in mesoporous films; the main one is a full alignment into a 'monocrystalline-like' mesophase. Only when the mesoporous channels show the same in-plane uniaxial orientation on a large length scale and are not only aligned in different directions parallel to the substrate do we achieve a fully mono-oriented porous phase. To achieve this, the surface can help a lot; if, in fact, the substrate is properly treated, for instance, by coating it with a mechanically rubbed polyimide film, the

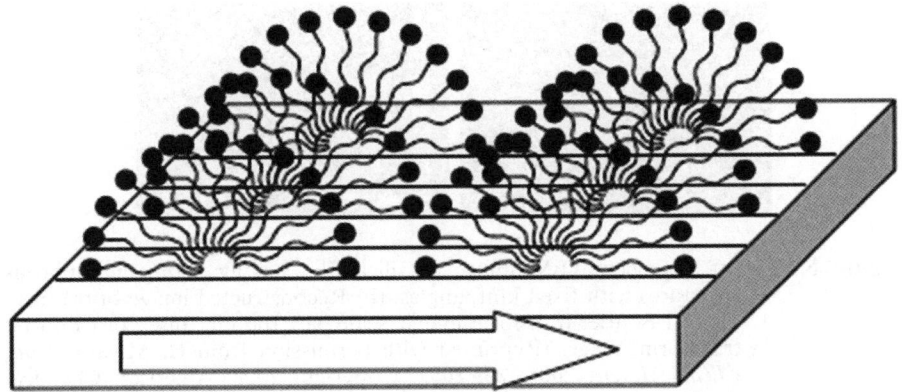

Figure 8.10 Schematic drawing of the alignment mechanism of the mesopores on a substrate coated with a rubbing-treated flexible polyimides. The arrow indicates the rubbing direction. (Reprinted with permission from H. Miyata, *Microporous Mesoporous Mater.*, 2007, **101**, 296. Copyright 2007, Elsevier.)

mesoporous 2D Hex films epitaxially grow and align following the traces imprinted on the substrate.[29] The surface properties guide the alignment process, and the channel orientation is governed by the interactions between the surface and surfactant molecules (**Figure 8.10**).[30] This method works well not only for achieving uniaxially aligned 2D Hex mesostructures but also single-crystalline 3D Hex mesostructures with full control of the in-plane arrangement of the spherical mesopores.

Relying only on the surface for orienting the pores is not the only method, and a different example is the application of a hot jet air flow to a droplet of precursor solution that has been deposited on the substrate; the mesochannels are oriented along the direction of the air flow.[31] This method gives the possibility of preparing multi-layered mesostructured films where every layer is oriented in a different direction. Some more 'exotic' in-plane orientations are even possible. The steric hindrance effects between the hydrophilic head of adsorbed surfactant molecules and silica oligomers on an aligned rubbing-treated polyimide film give two distinct alignment directions with a zigzag porous structure and a fixed kinked angle (**Figure 8.11**).[32] These types of alignments are not only an exercise of self-assembly but have some important practical applications. Mesoporous silica films with kink structure have been used to reduce the translocation speed of DNA during its detection.[33] In comparison with straight porous structures of similar dimensions, kinked mesopores have shown up to a five-fold

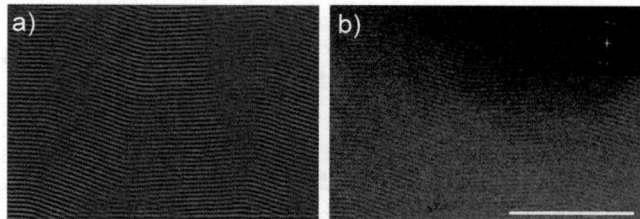

Figure 8.11   (a) Top view TEM image of silica film having zigzag mesoporous structures with fixed kink angles. (b) Reconstructed image of (a) from the fast Fourier transform image, scale bar: 100 nm. Inset: fast Fourier transform image. (Reprinted with permission from H. Miyata *et al.*, *Chem. Mater.*, 2008, **20**, 1082. Copyright 2008, American Chemical Society.)

reduction in translation velocity of DNA, which is one of the critical issues in DNA sequencing. These are examples of in-plane orientation, but obtaining ordered arrays of channels that are orthogonally oriented with respect to the substrate is even more difficult. Electrochemically assisted self-assembly[34] and nanometer-scale epitaxial growth[35] have been developed for aligning the pores; both the methods allow for achieving vertical alignment of oriented hexagonal channels in a vertical direction with respect to the substrate. This is not limited to perfectly aligned vertical channels, and tilted mesochannels have been obtained by modifying the substrate with PEO−PPO copolymers for achieving a chemically neutral surface that promotes alignment[36] (**Figure 8.12**).

8.8   A LIVING STRUCTURE

The 'soft' nature of the template in mesostructured films gives some specific properties to the material; until the removal of the micelles the material has a hybrid structure with an organic template surrounded by an inorganic (or organic−inorganic) network. This makes the material quite responsive to external stimuli, for instance the micelles can be easily swollen by water absorption while the oxide network is not yet fully condensed; until the material structure is frozen by thermal treatment it remains in a state that has been defined as a tunable state. This specificity causes the overall mesostructured film to behave as a soft material whose structure can still be rearranged to a large extent by external intervention. The tunable state of mesoporous films has been discovered by *in situ* analysis and then widely applied for controlling the meso-organization during film processing. Modulating the humidity and ethanol vapor within the deposition room is the simplest way to

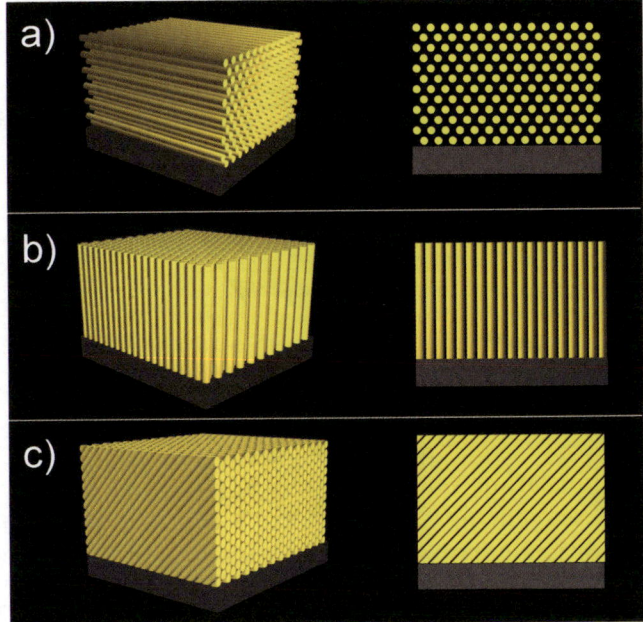

Figure 8.12 Possible orientations of 2D hexagonal pore structures in mesoporous films with respect to the substrate.

induce a reversible transition of the mesophase if the oxide structure is still sufficiently flexible to allow reorganization of the templating micelles into a different ordered structure. During film deposition the relative humidity (RH) in the deposition room dictates the type of organization, a change in RH induces micelle swelling and reorganization into a different phase (**Figure 8.13**). It is interesting that the material has a surprising capability for responding to different types of external stimuli, and changes in the mesophase have been observed in deposited films exposed to UV light and X-rays.

In such soft structures the thermal treatment also produces different effects, the simplest being the condensation and removal of the templating surfactant. In turn, the thermal process produces shrinkage of the films in the direction normal to the substrate with distortion of the pore shape, generally from spherical to ellipsoidal, and also a change in the pore organization. The cell parameters, even if phase transition of the mesostructure is not observed, can change upon shrinkage. In several cases, however, a true rearrangement of the organization is observed, which has to follow in any case the symmetry rules with only some transitions allowed. Some of the transitions that

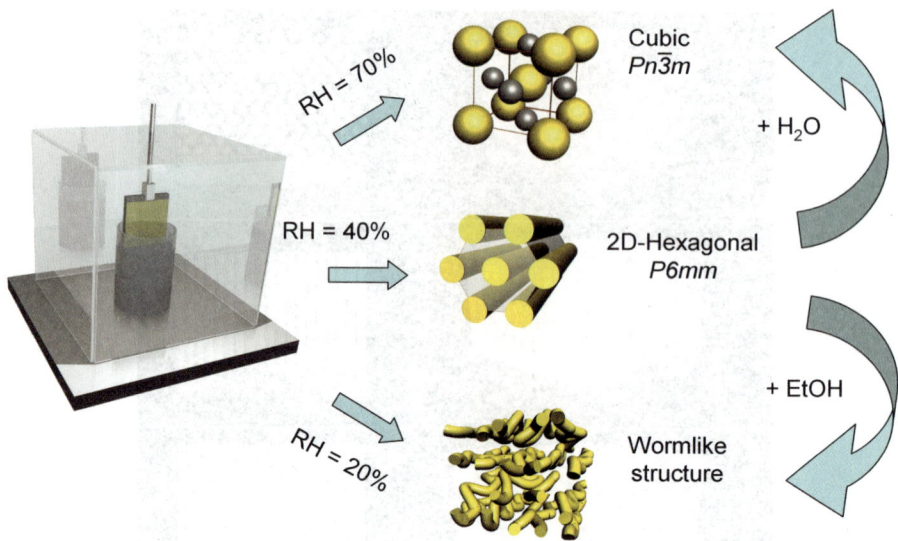

Figure 8.13   Mesoporous phases obtained at different RH values. Phase transitions are induced by water and ethanol vapor pressure variation when systems are in the tunable state.

have been reported are from oriented (10) planar hexagonal *p6mm* to an oriented (10) planar rectangular *c2mm* mesophase; from an oriented (110) body-centred cubic $Im\bar{3}m$ to an oriented (010) orthorhombic

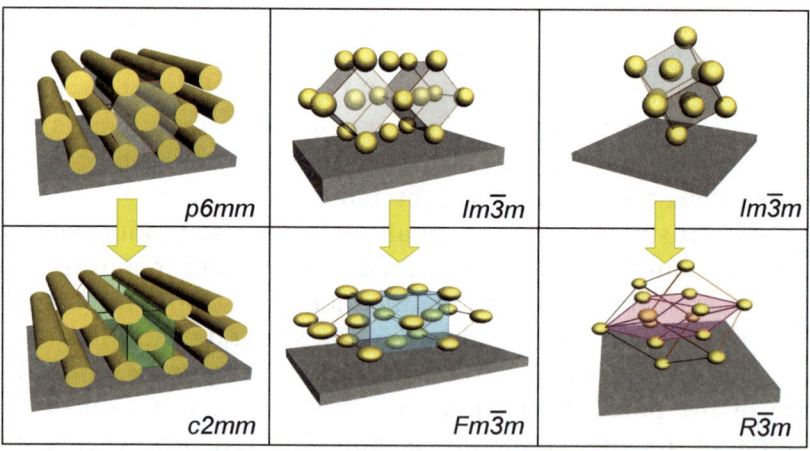

Figure 8.14   Allowed mesophase transitions in ordered mesoporous silica films. (Reprinted with permission from P. Innocenzi *et al.*, *Chem. Mater.*, 2009, **21**, 2555. Copyright 2009, American Chemical Society.)

Fmmm, and from an oriented (111) face-centred cubic $Fm\bar{3}m$ to an oriented (111) rhombohedral $R\bar{3}m$ phase (**Figure 8.14**).

REFERENCES

1. This is the definition of porosity given by the International Union of Pure and Applied Chemistry (IUPAC).
2. D. W. Breck, *Zeolite Molecular Sieves: Structure, Chemistry and Use*, Wiley Interscience, New York, 1974.
3. J. S. Beck, J. C. Vartuli, W. J. Roth, M. E. Leonowicz, C. T. Kresge, K. D. Schmitt, C. T. W. Chu, D. H. Olson, E. W. Sheppard, S. B. McCullen, J. B. Higgins and J. L. Schlenker, *J. Am. Chem. Soc.*, 1992, **114**, 10834.
4. C. T. Kresge, M. E. Leonowicz, W. J. Roth, J. C. Vartuli and J. S. Beck, *Nature*, 1992, **359**, 710.
5. N. K. Raman, M. T. Anderson and C. J. Brinker, *Chem. Mater.*, 1996, **8**, 1682.
6. J. N. Israelachvili, D. J. Mitchell and B. W. Ninham, *J. Chem. Soc., Faraday Trans. 2*, 1976, **72**, 1527.
7. N. Hadjichristidis, S. Pispas and G. A. Floudas, *Block copolymers. Synthetic strategies, physical properties, and applications*, Wiley-Interscience, Chichester 2003, 203.
8. *Block Copolymers in Nanoscience*, ed. M. Lazzari, G. Liu and S Lecommandoux, Wiley-VCH, Weinheim, 2006, 291.
9. S. Förster and T. Plantenberg, *Angew. Chem., Int. Ed.*, 2002, **41**, 688.
10. G. J. A. A. Soler-Illia, E. L. Crepaldi, D. Grosso and C. Sanchez, *Curr. Opin. Colloid Interface Sci.*, 2003, **8**, 109.
11. The definition of 'soft matter' is applied to different physical states of condensed matter, which are easily deformed by thermal stresses or thermal fluctuations. This matter includes for instance colloids, gels, polymers and foams.
12. C. J. Brinker and G. W. Scherrer, *Sol−Gel Science, The Physics and Chemistry of Sol−Gel Processing*, Academic Press, San Diego, 1990.
13. G. J. A. A. Soler-Illiia and O. Azzaroni, *Chem. Soc. Rev.*, 2011, **40**, 1107.
14. L. Nicole, C. Boissière, D. Grosso, A. Quach and C. Sanchez, *J. Mater. Chem.*, 2005, **15**, 3598.
15. C. Sanchez, C. Boissiere, D. Grosso, C. Laberty and L. Nicole, *Chem. Mater.*, 2008, **20**, 682.
16. G. J. A. A. Soler-Illia and P. Innocenzi, *Chem. − Eur. J.*, 2006, **12**, 4478.
17. Q. Huo, D. I. Margolese, U. Ciesla, D. G. Demuth, P. Feng, T. E. Gier, P. Sieger, A. Firouzi, B. F. Chmelka, F. Schzth and G. D. Stucky, *Chem. Mater.*, 1994, **6**, 1176.
18. C. J. Brinker, Y. Lu, A. Sellinger and H. Fan, *Adv. Mater.*, 1999, **11**, 579.
19. D. Grosso, P. A. Albouy, H. Amenitsch, A. R. Balkenende and F. Babonneau, *Mater. Res. Soc. Symp. Proc.*, 2002, **628**, CC6.17.1.

20. D. Grosso, A. R. Balkenende, P. A. Albouy, A. Ayral, H. Amenitsch and F. Babonneau, *Chem. Mater.*, 2001, **13**, 1848.
21. D. A. Doshi, A. Gibaud, V. Goletto, M. Lu, H. Gerung, B. Ocko, S. M. Han and C. J. Brinker, *J. Am. Chem. Soc.*, 2003, **125**, 11646.
22. D. Grosso, F. Babonneau, P. A. Albouy, H. Amenitsch, A. R. Balkenende, A. Brunet-Bruneau and J. Rivory, *Chem. Mater.*, 2002, **14**, 931.
23. D. Grosso, F. Cagnol, G. J. A. A. Soler-Illia, E. L. Crepaldi, H. Amenitsch, A. Brunet-Bruneau, A. Bougeois and C. Sanchez, *Adv. Funct. Mater.*, 2004, **14**, 309.
24. P. C. A. Alberius, K. L. Frindell, R. C. Hayward, E. J. Kramer, G. D. Stucky and B. F. Chmelka, *Chem. Mater*, 2002, **14**, 3284.
25. P. Innocenzi, L. Malfatti, T. Kidchob and P. Falcaro, *Chem. Mater.*, 2009, **21**, 2555.
26. D. Grosso, C. Boissière and L. Nicole, *J. Sol−Gel Sci. Technol.*, 2006, **40**, 141.
27. P. Falcaro, S. Costacurta, G. Mattei, H. Amenitsch, A. Marcelli, M. C. Guidi, M. Piccinini, A. Nucara, L. Malfatti, T. Kidchob and P. Innocenzi, *J. Am. Chem. Soc.*, 2005, **127**, 3838.
28. V. N. Urade, T. C. Wei, M. P. Tate, J. D. Kowalski and H. W. Hillhouse, *Chem. Mater.*, 2007, **19**, 768.
29. H. Miyata, T. Suzuki, A. Fukuoka, T. Sawada, M. Watanabe, T. Noma, K. Takada, T. Mukaide and K. Kuroda, *Nature*, 2004, **3**, 651.
30. H. Miyata, *Microporous Mesoporous Mater.*, 2007, **101**, 296.
31. B. Su and Q. Lu, *J. Am. Chem. Soc.*, 2008, **130**, 14356.
32. H. Miyata, T. Suzuki, M. Watanabe, T. Noma, K. Takada, T. Mukaide and K. Kuroda, *Chem. Mater.*, 2008, **20**, 1082.
33. Z. Chen, Y. Jiang, D. R. Dunphy, D. P. Adams, C. Hodges, N. Liu, N. Zhang, G. Xomeritakis, X. Jin, N. R. Aluru, S. J. Gaik, H. W. Hillhouse and C. J. Brinker, *Nat. Mater.*, 2010, **9**, 667.
34. Y. Yamauchi, M. Sawada, A. Sugiyama, T. Osaka, Y. Sakka and K. Kuroda, *J. Mater. Chem.*, 2006, **16**, 3693.
35. E. K. Richman, T. Brezesinski and S. H. Tolbert, *Nat. Mater.*, 2008, **7**, 712.
36. V. R. Koganti, D. Dunphy, V. Gowrishankar, M. D. McGehee, X. Li, J. Wang and S. E. Rankin, *Nano Lett.*, 2006, **6**, 2567.

CHAPTER 9

Towards the Complex Organization of Matter: Hierarchical Porosity

In the previous chapter we have shown how self-assembly can push a system towards the formation of long-range organized structures. Besides the direct outcomes, the most fascinating implication suggested by this approach is the possibility to program a random liquid system to self-order under controlled evaporation conditions. In other words: can we teach matter how to organize itself to produce functional devices independently? Despite the recent breakthroughs in the field of self-assembly, this task still remains an ambitious goal. However, the scientific community is trying to improve the complexity of the bottom-up-made systems by increasing the number of self-assembly processes occurring during material formation. This strategy leads to the so-called *hierarchical materials*.

9.1 POROUS HIERARCHICAL MATERIALS

The Oxford English Dictionary gives the following definition of *hierarchy*: 'a ranking system ordered according to status or authority'.[1] This definition, however, sounds slightly unsuited for describing the organization of matter, because it would be hard to establish a 'status' or an 'authority' for different lifeless structures. However, there is no doubt that several systems in nature appear to be organized according to a precise architecture. If we think about wood, for example, we should recognize that its structure exhibits a defined 'anatomy', from macro- down to the nanoscale (**Figure 9.1**).[2] In hardwood, for example, we find at first some pipe-like structures (vessels) devoted to lymph transportation with a diameter in the range of 500 μm, then smaller

Water Droplets to Nanotechnology: A Journey Through Self-Assembly
By Plinio Innocenzi, Luca Malfatti and Paolo Falcaro
© P. Innocenzi, L. Malfatti and P. Falcaro 2013
Published by the Royal Society of Chemistry, www.rsc.org

Figure 9.1   The hierarchical structure of wood. (Reprinted with permission from ref. 2.)

tube-like cells (tracheids) oriented in the stem (20 to 50 μm). Moreover, the wood's cell walls show a composite structure made of cellulose fibres embedded in a matrix made of non-cellulosic polysaccharides (hemicelluloses), lignin, and inorganic compounds. The fibres have dimensions from 50 to 200 nm with a core–shell structure. The inner part is crystalline and the outer part is amorphous.[3] So, what is the 'status' that is considered for the organization of matter in the wood? The answer lies in the different roles played by each single structure at every length scale. Lymph transportation, water absorption or release, mechanical strength and stiffness are different functions that are carried out by specialized structures organized according to a specific project and refined by millions of years of evolution.

This chapter deals with hierarchical porosity. We have restricted the field to thin films that show some peculiarities and specific properties; this is in accordance with the choice that we have made for describing self-assembly in thin films (see Chapter **8**). Hierarchical porosity in films is defined as 'a 3D arrangement of well-defined pores of different sizes, the smaller ones being located in the walls between the larger pores, thereby also establishing the connectivity'.[4] Multimodal pore distribution with a defined architecture is envisaged to become a multifunctional system where specific properties and roles are assigned to different porosities of the materials according to their size and shape. Most research on hierarchical porous films has been dedicated to silica-based systems,[5] but a few examples of non-siliceous-based materials have been also reported.

9.2 AN EASY APPROACH: EVAPORATION-INDUCED SELF-ASSEMBLY ON PRE-PATTERNED SUBSTRATES

A relatively simple process for forming hierarchical porous structures in films is by coupling the self-assembly through the evaporation process described previously with other techniques that enable organization of porosity at the macroscale. The first process allows for achieving materials with porosity on the mesoscale, within a typical range of 2–20 nm, while macropores will be larger than 50 nm. A possible route for combining the two ranges of porosity within the same film is to use a sacrificial macroporous framework; preformed membranes made of cellulose acetate or polyamide can be used as a macroporous scaffold for mesopores.[6] The membranes are, at first, impregnated with a mesoporous templating solution, then deposited on a substrate, dried and calcined in an oven. After the thermal treatment, which removes the organic scaffold, a hierarchical porous silica film showing both meso and macropores is obtained. In addition, the film has a remarkable thickness of several micrometers and a high surface area estimated from about 400 up to 800 m^2 g^{-1} according to the choice of the mesopore templating agent.

The macroporous scaffold can be also fabricated *via* lithography, using two-photon polymerization (2PP) to produce micrometer-sized scaffolds for mesoporous films.[7] 2PP is a lithographic technique that enables selective 3D polymerization by using a femtosecond pulsed laser. The process starts with the micro-fabrication[8] of a resin layer cast on a glass substrate by 2PP producing a 3D wood-pile macroporous scaffold, and then the polymerized resin is coated by a mesoporous titania or silica film through dip-coating. After deposition, thermal calcination is used to remove the macroporous polymeric framework and, at the same time, the templating agent forms the mesopores. The final material, therefore, is a mesoporous inorganic oxide shaped as a replica of the macroporous organic scaffold. The whole process is illustrated in **Figure 9.2**.

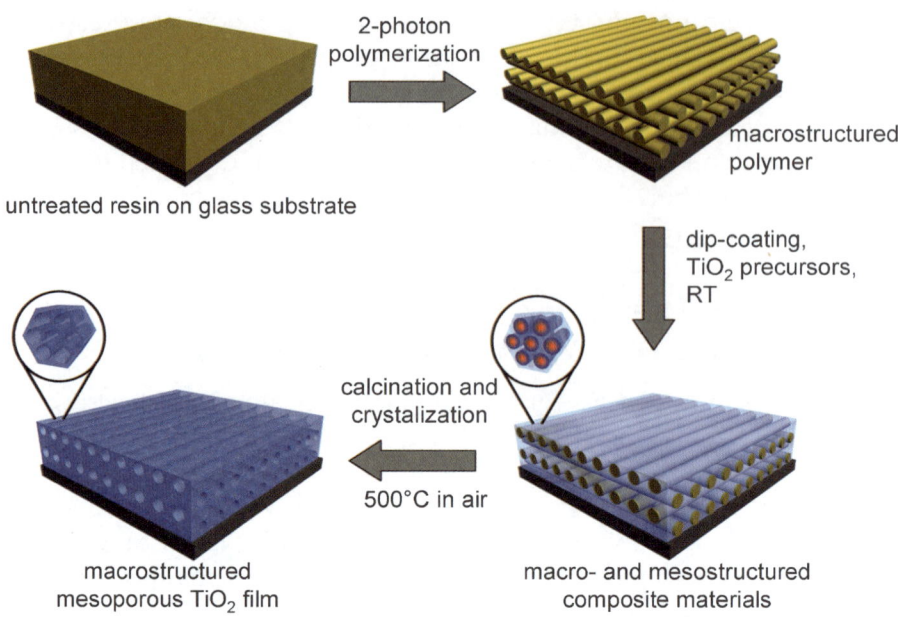

Figure 9.2   Synthesis of mesostructured silica-based hybrids with ordered macropores. (Adapted with permission from P. Innocenzi *et al.*, *Chem. Mater.*,
2011, **23**, 2501. Copyright 2011, American Chemical Society.)

Texture at the macroscale can be obtained not only by a top-down
approach, such as 2PP, but also by a bottom-up technique, such as
controlled phase separation of a sacrificial substrate. An example is the
preparation of a macroporous titania coating through the phase-
separation of poly(ethyleneglycol) (PEG) during film deposition.[9] The
procedure allows for obtaining textured films with vertical pores ranging
from 100 nm to 1 μm in diameter.[10] Afterwards, the macroporous
scaffold has been coated with mesoporous silica producing a multi-
layered structure. Finally, the macroporous titania film is selectively
removed through an etching treatment by exposing the bilayer film to an
acidic peroxide solution. The whole procedure, therefore, allows for
obtaining a titanium-free macro-textured mesoporous silica film that
would be particularly suitable for a biological application, such as the
surface coating of bone implant materials.

Mesoporous films with sub-micrometer bump-like structures have
also been prepared by pre-patterning of the substrate with sacrificial
carbonaceous islands.[11] This method requires the use of a sucrose-
surfactant under acidic conditions. Upon drying, dehydration of the
saccharide leads to the formation of the carbon-based disk-like islands.
After deposition of a mesoporous silica coating on the patterned

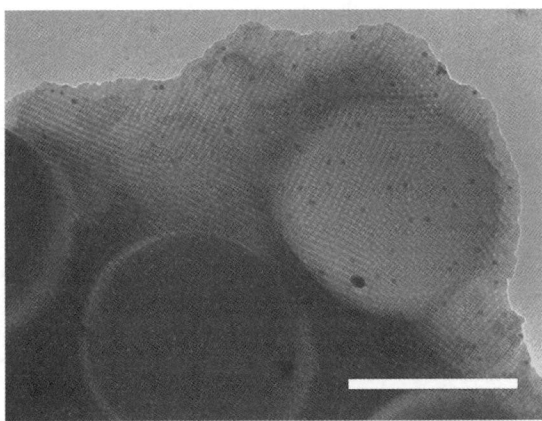

Figure 9.3   TEM images showing the 'nano-vaults' in mesoporous silica films. (Reprinted with permission from ref. 10.)

substrate, the carbon structures can be removed leaving a homogenous inorganic film showing 'vaults' made of mesoporous silica with a diameter of around 1 μm (**Figure 9.3**).

9.3   THE HIERARCHICAL ASSEMBLY OF POROUS PARTICLES

The coatings made of stacked porous particles offer interesting advantages when compared to hierarchical porous films obtained from the liquid phases. In particular, the particle dimension and the stacking density allow for controlling two main properties: the macroporosity, induced by bottle-necks among particles, and the surface roughness, which is tuned by the nanoparticles' size and shape. This strategy has been adopted for preparing hydrophobic and antireflective coatings by the assembly of mesoporous silica nanoparticles with non-spherical shapes.[12] In the first step, mesoporous silica particles with an average size of 70–90 nm are fabricated by combining two templating agents, an ionic surfactant (cetyltrimethylammonium bromide, CTAB) and a hydrophobic drug (ibuprofen, IBU).[13] The former is responsible for the mesopores formation in the nanoparticles while the latter allows for tuning the morphology and structure of the silica nanospheres. Under basic conditions, IBU is negatively charged and has a strong electrostatic interaction with the CTA^+, therefore the drug can be considered as a co-surfactant for the synthesis of non-spherical particles. Once prepared, the nano-objects are deposited on a glass substrate by using a so-called layer-by-layer (LbL) assembly. This technique allows for preparing homogeneous thin films with precise control over film properties by using nanoparticles paired with polyelectrolytes. After calcination, the polyelectrolytes are removed by thermal degradation leaving a fully inorganic hierarchical structure. The high degree of porosity of the materials changes the refractive index in the coatings and this is responsible for the antireflective property while the controlled surface roughness makes the film hydrophobic by the so-called 'lotus effect'.[14]

Hierarchical porous nanoparticles represent an alternative to monomodal porous particles for film preparation.[15] After thermal treatment, the particles have a size of 150–220 nm with a bimodal distribution of porosity, macropores of 5–30 nm, and mesopores of 2–3 nm in diameter. The smaller pores are generally ordered with a topological structure while the pores of larger dimensions remain randomly distributed. The coatings obtained from nanoparticle stacking have superhydrophilic properties because of the extremely large surface area with an abundance of 'nanovoids' and 'nano-reservoirs'. However, after surface modification with 1H,1H,2H,2H-perfluorooctyltriethoxysilane the coatings become superhydrophobic with extremely large water contact angles ($\geqslant 150°$).

Another route, which is a little more complex, is the encapsulation of functionalized mesoporous silica nanoparticles into a hierarchical porous matrix. **Figure 9.4** shows a schematic that outlines the procedure. In this case the final target is the synthesis of a chemical sensor with hierarchical structure where the larger pores contain the silica nanoparticles and the mesopores ensure the interconnectivity among the

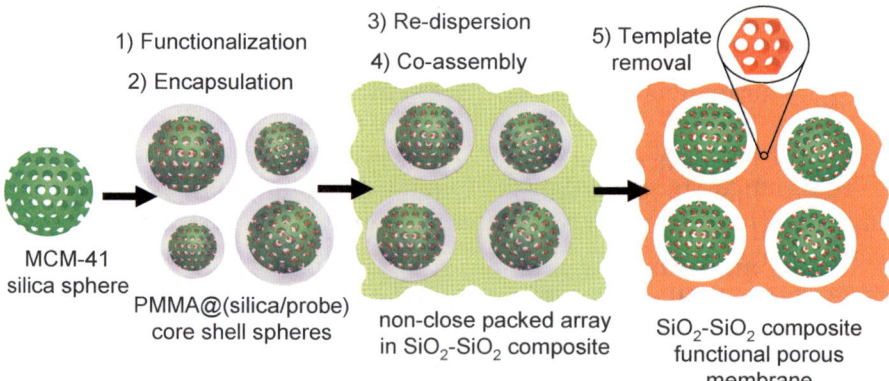

Figure 9.4   Drawing of the fabrication of a hierarchical porous nanocomposite with embedded functional mesoporous silica nanoparticles. (Adapted with permission from F. Li *et al.*, *Chem. Mater.*, 2010, **22**, 3790. Copyright 2010, American Chemical Society.)

macropores.[16] The preparation of functional hierarchical materials is achieved in different steps. At first the mesoporous nanoparticles are functionalized with a chromophore to enable them to detect the presence of Cd^{2+} ions in solution by a color change. Then, the particles are coated with a protective shell of poly(methylmethacrylate) (PMMA) and used as templates for macropores in a precursor solution for the preparation of a mesoporous silica film through EISA. The subsequent thermal treatment allows for removing the surfactant from the mesopore and the polymeric shell of the silica particles. The final coating shows macropores partially filled by silica nanoparticles with a smaller dimension. Such macropores are separated by mesoporous walls, which ensure liquid permeability. The material works well as an optical sensor of Cd cations and shows a sharp color change induced by the heavy metal ions in solution.

9.4   ONE-POT SYNTHESIS WITH PREFORMED TEMPLATES

The most exploited method for obtaining hierarchical porous films combines the evaporation self-assembly route that achieves mesoporosity with the addition of preformed templates having larger dimensions with respect to the mesoscale. Usually, polymeric beads or microparticles made of polystyrene (PS) or poly(methylmethacrylate) (PMMA) are used to obtain macropores, but this approach has a main issue to be tackled. Direct one-pot addition of the micron-sized beads generally leads to aggregation of the macro templates and causes a patchy dispersion of the larger pores in the hierarchical materials. Therefore, the spatial distribution of the preformed templates is one of the most critical constraints that this synthesis has to face. Fortunately, film formation from a liquid phase is a very flexible technique, which allows obtaining similar results by using different combinations of multiple parameters such as pH, concentration of sol−gel precursors, deposition technique (dip- or spin-coating) and type of solvent. Adjusting the concentration and the mixture of ethanol and water used for film deposition is an interesting method to synthesize hierarchical porous silica films with very low refractive index.[17] Increasing amounts of a colloidal solution, made of water, surfactant and 70 nm sized polymeric nanoparticles were added to a sol−gel solution containing triblock copolymer surfactant, tetraethylorthosilicate, methyltriethoxysilane, ethanol and an acidic catalyst. The solution produces mesoporous films over a wide concentration range of nanoparticle colloidal solution, without affecting the self-assembly process and the degree of mesopore order. The polymeric particles remain well dispersed in the films producing a homogeneous hierarchical porous structure after thermal calcination. It is interesting to note that the uniaxial constraint of the film produces on both meso and macropores a uniaxial shrinkage as a function of temperature so that the pores in the calcined coatings show an elliptical shape (**Figure 9.5**). The porous films prepared by this method also exhibit a peculiar optical feature, which is an ultralow refractive index that is lower than the value measured for the monomodal mesoporous film. This property is attributed to the large amount of air, *i.e.* pores, contained in the coatings.

A porous architecture with hierarchical organization is crucial for many applications where a high surface area has to be easily accessed by a liquid phase. This is the case for dye-sensitized solar cell (DSSC) devices where an electrolytic liquid mediator is generally required to close the circuit and produce photocurrent.[18] An example of this type is the use of poly(dimethyl siloxane)-block-methyl methacrylate poly (ethylene oxide) (PDMSb-MA(PEO)) as a structure directing agent for small pores and 1 μm sized PMMA microsphere to template larger pores in hierarchical porous titania films.[19] The advantage of this method is the removal of both templates by chemical dissolution with acetic acid, instead of thermal calcination, as shown in **Figure 9.6**. The films show a disordered network of pores with an average dimension of

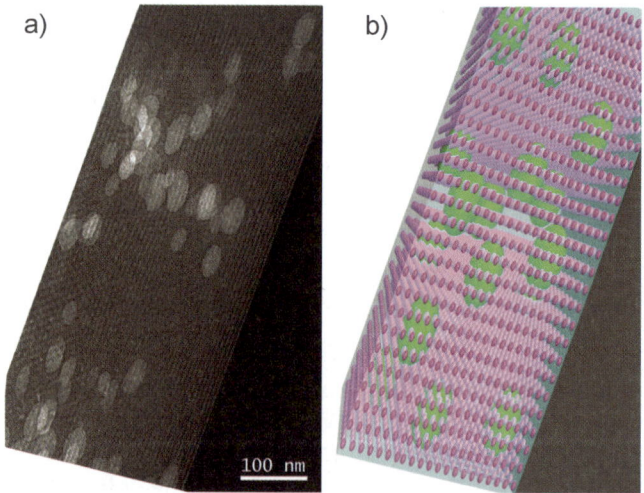

a)   b)

Figure 9.5   TEM image and 3D rendering of hierarchical porous silica with elliptic pore nanoparticles. (Adapted with permission from P. Falcaro *et al.*, *Chem. Mater.*, 2009, **21**, 2055. Copyright 2009, American Chemical Society.)

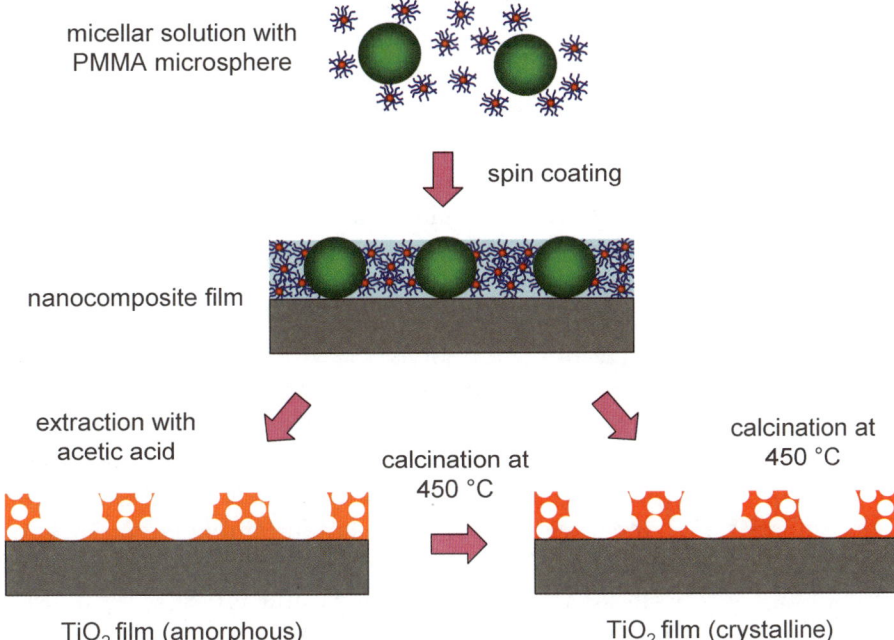

Figure 9.6 Illustration of the method use to produce hierarchical porous titania films. (Adapted with permission from G. Kaune *et al.*, *ACS Appl. Mater. Interfaces*, 2009, **1**, 2862. Copyright 2009, American Chemical Society.)

52 nm and macropores with a crater-like surface depression about 1 µm in diameter. In addition, some porous features with size $\approx$ 180 nm are obtained from liquid−liquid phase separation that occurs during film deposition.[20] These hierarchical titania coatings perform three functional properties according to their organization on multiple length scales. In fact, at the micrometer scale the crater-like structures reduce light reflection, at an intermediate scale macropores improve infiltration of electrolytic mediator and at a nanometer scale the smaller porous structure increases the electric charge separation produced by a photovoltaic effect.

The simultaneous use of two templates has been employed to improve the photocatalytic activity in hierarchical porous titania films.[21] A combination of self-assembly and direct templating with 100 nm sized PS spheres produces titania films with meso–macroporosity. Tuning the concentration of the block copolymer, which is also responsible for mesopore formation, minimizes the bead aggregation although some clustering of PS beads is unavoidable. Larger pores, which form by the merging of PS beads, are also important for keeping the mesostructure organization after film processing and allowing a better permeability of the whole pore surface. After irradiation with UV light, the photodegradation properties of the hierarchical titania coating improve because of the effective diffusion in the PS-derived macropores.

The hierarchy of the pores is not restricted to two orders; the possibility of increasing the level of complexity depends of course on the ability of combining different templates in the appropriate synthesis route. The simultaneous use of three different templating agents to induce trimodal porosity can give pores of different sizes.[22] A combination of poly(x-hydroxypoly(ethylene-cobutylene)-co-poly(ethylene oxide) (KLE-type block copolymer) with an ionic liquid, 1-hexadecyl-3-methylimidazolium chloride, has been used to form smaller and larger mesopores with an average size of 3 and 14 nm, respectively. The smaller pores show a worm-like structure among the larger pores that, in contrast, appear well ordered in a close-packed cubic fashion. Very interestingly, the presence of ionic liquid-templated mesopores affects, to some extent, the 3D organization of the block-copolymer-templated pores as it screens the interactions and inhibits the mutual correlation among the close packed layers of spherical mesopores. Besides these templates, monodispersed PMMA beads with an average diameter of $\approx$ 150 nm have also been found to be suitable macropore structuring agents. The shrinkage occurring during calcination reduces the macropore dimension to about 125 nm and leaves a hierarchically structured 3D pore system. Hierarchical films are expected to have a better performance in specific applications because of the higher interpore connectivity and higher surface area. The better electro-chemical response of ferrocene-functionalized hierarchical films has experimentally confirmed this assumption.

9.5 OPAL INFILTRATION AND MICRO-MOULDING

Instead of being directly added to the precursor sol, polymeric beads and microparticles can be pre-processed to form long-range ordered structures. These objects can be stacked in 2D layers of particles or in 3D opal arrangements. Once the beads have been organized, the empty spaces in the periodic structures can be filled with templating sol−gel solution. After thermal removal of the meso- and macro-templates, the inorganic framework shows an organized porosity on both length scales.

By following this approach, hierarchical porous silica films have been obtained using electro-deposition of mesoporous oxide on a preformed multilayer array of PS particles.[23] The electro-assisted deposition of templating sol−gel solution allows for achieving mesoporous coatings of silica with cylindrical pores aligned perpendicularly to the substrate.[24] This pore symmetry is of extreme importance for many emerging applications such as DSSC (*vide supra*), nanofluidicis or, more generally, processes involving mass transportation. The material is prepared by deposition on a conductive glass substrate of a stack of PS particles with a dimension of 100 nm by dip-coating at slow withdrawal speeds. In the second step the particles are coated with a mesoporous oxide by electrochemical deposition. The mesoporous silica clusters penetrate into the voids among the polymeric particles and gradually form a thick inorganic wall. The formation of a robust hierarchical porous structure depends, therefore, by the time used for electro-deposition, longer deposition times lead to thicker and more oriented mesoporous films with pore channels perpendicular to the glass substrate. **Figure 9.7** shows the morphology of the different coatings after thermal treatment at 550 °C.

A higher degree of organization of the macropores in hierarchical structures can be obtained by the use of opal scaffolds made by polymeric microspheres. Usually opals are produced by convective self-assembly (see Chapter **4**) or dip-coating (see Chapter **6**) of colloidal solutions. Once the periodic structure has been built, a templating solution of mesoporous material can be infiltrated among the microparticles to form the inorganic network. A final thermal treatment allows for removing the pore templates and leaves an open accessible bimodal porosity. Using this approach, hierarchical porous materials have been fabricated as a sensor for chemical compounds.[25] Silica films with macro and mesoporous structures using PS spheres and an ionic surfactant (CTAB) as structure-directing agents have been prepared. The macropores and mesopores have a size of 450 and 2 nm, respectively. The doping of the porous structure with porphyrin also allows for fabricating a fluorescent chemical sensor to detect trace amounts of explosives in the vapor state, such as 2,4,6-trinitrotoluene, 2,4-dinitrotoluene and nitrobenzene.

An alternative method to fabricate bimodal porosity[26] is the infiltration of a solution containing inorganic sol−gel precursors and surfactant in polymeric opals. In this case, specifically designed hybrid silica precursors bearing alkyl chains with different lengths have to be employed. Upon spin coating the opal structures the hybrid precursors self-organize forming 2D Hex or lamellar mesophases inside the voids of the ordered macro templates. The whole process is shown in **Figure 9.8**. After thermal treatment, the alkyl chains can be removed from the

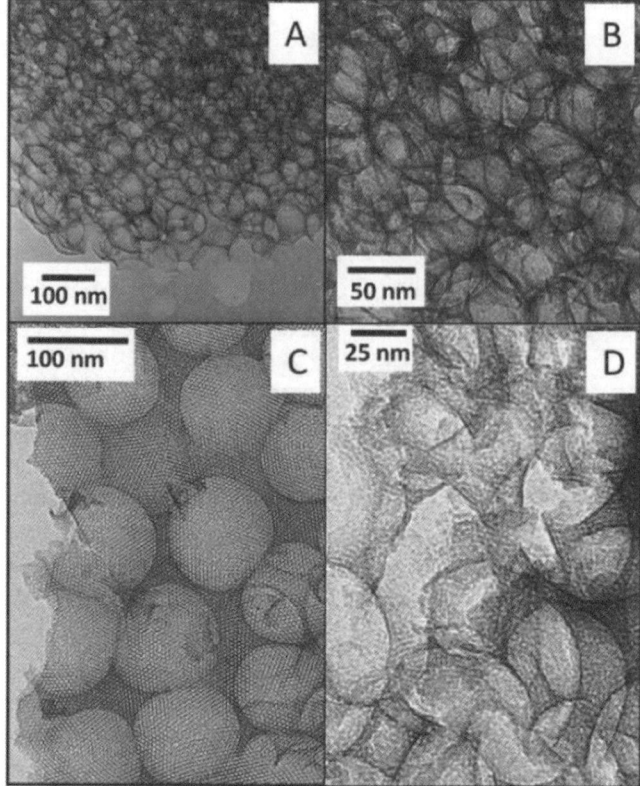

Figure 9.7 TEM pictures of hierarchical silica film fragments obtained by electrochemical generation under different conditions. (Reprinted with permission from M. Etienne *et al.*, *Chem. Mater.*, 2010, **22**, 3426. Copyright 2010, American Chemical Society.)

matrix leaving a hierarchical porous structure with macropores of 420 nm and micropores of 1.4 nm.

A combination of latex sphere self-assembly, EISA and micromoulding is at the basis of a smart approach to produce hierarchically ordered porous oxides such as silica, titania and niobia.[27] Micromoulding is a technique that uses capillary forces to fill patterned channels in a mould.[28] The mould is then used to stamp a replica of the patterned channels on a substrate by an applied pressure. A sol−gel solution of precursors containing both block copolymers as the template for the mesopores, and latex microspheres as the macropore-directing agent. As clearly shown in the SEM images (**Figure 9.9**), this technique allows for fabricating hierarchical inorganic structures with discrete length scales of 10, 100 and 1000 nm in a single matrix. Self-assembly *via* evaporation is responsible for pore templating on the scale of 10 nm,

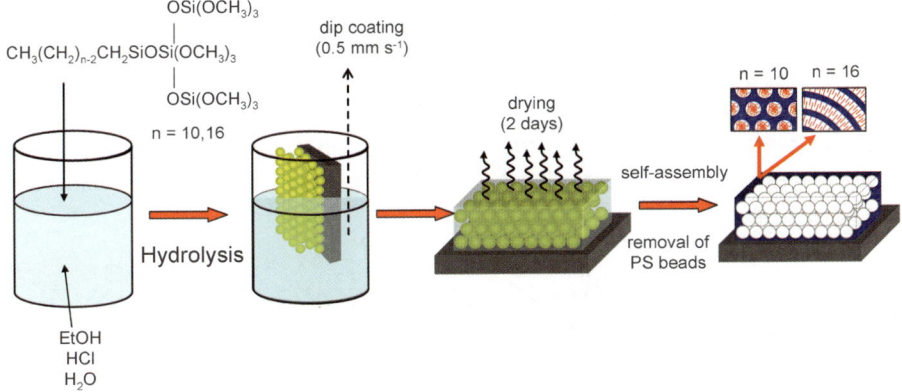

Figure 9.8 General strategy to obtain hierarchical porous silica from mesostructured silica-based hybrids with ordered macropores. (Adapted with permission from M. Sakurai *et al.*, *Langmuir*, 2007, **23**, 10788. Copyright 2007, American Chemical Society.)

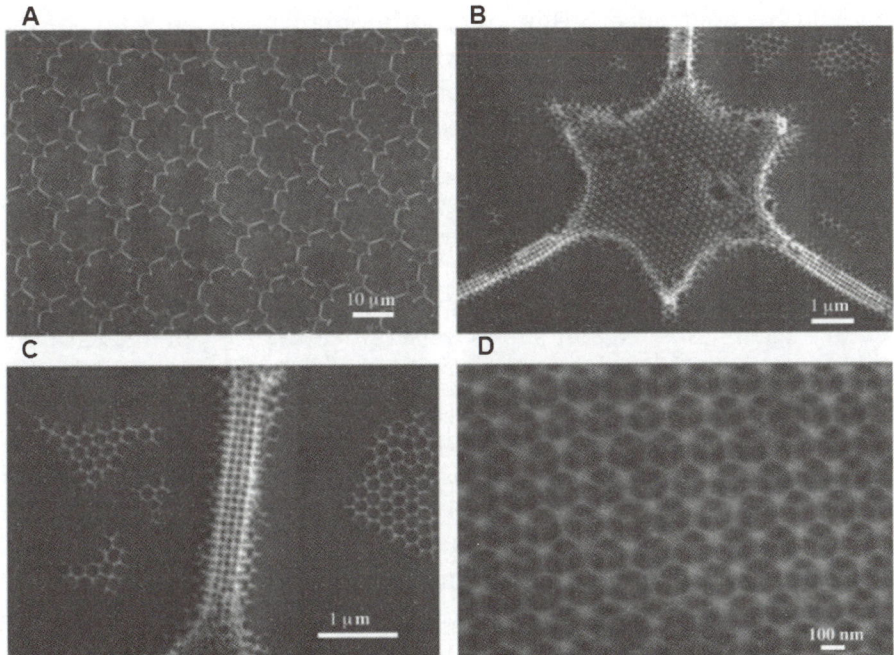

Figure 9.9 SEM images, at different magnifications, of hierarchically ordered porous silica displaying organization over three different length scales. (Reprinted with permission from: P. Yang *et al.*, *Science*, 1998, **282**, 2244. Copyright 1998, AASS.)

latex sphere self-assembly for organization at 100 nm and the micro-moulding lithography for the fabrication at the micrometer scale. The mesopores have a cubic or hexagonal long-range-ordered structure while the latex spheres are stacked in close-packed arrays. The organization of the macro templates is driven by the capillary forces among the microspheres and convection motion due to solvent evaporation.

9.6 ONE-POT APPROACH WITH *IN SITU* FORMATION OF MULTIPLE TEMPLATES

We have described several strategies that rely on preformed objects or scaffolds for obtaining hierarchical porous films. However, the most challenging route is the synthesis of hierarchical porous materials by controlling *in situ* the formation and self-assembly of multiple templates of different dimensions. How is it possible to do this? There are several possibilities, some quite elegant, that have been explored; clearly they should be based on chemical−physical phenomena that happen simultaneously with film processing. Outgassing, micro-phase separation and the controlled precipitation of inorganic salts are some of the processes that have been coupled with self-assembly through evaporation with a template to produce bimodal porosity in films. Although these techniques do not result in organization of the pores at both length scales they appear quite attractive because they allow for easy and fast film preparation by using commonly available reagents.

A first example is that of cellulose nitrate and block copolymers or ionic surfactant for producing hierarchical porous silica.[29] Cellulose nitrate (CN) is an explosive compound that was discovered over a century ago.[30] This reagent shows a useful feature for pore generation. It has a lower temperature of deflagration compared to combustion. This means that the compound prefers to decompose instead of combusting, which requires an oxidizer (such as molecular oxygen). If the gas obtained during decomposition is produced while the silica matrix retains relatively low viscosity, which is actually the case during film drying, a macroporous structure is formed in the final material. This method has allowed for deposition by spin-coating bimodal porous silica films with a thickness of 500 nm and surface area from 500 up to 800 m^2 g^{-1}. **Figure 9.10** shows the inner morphology of the films. The macropores that are generated by CN outgassing are in the range of 15 and 20 nm whilst the mesopores, produced by surfactant self-assembly, have a size of around 5 nm. Interestingly, the role of the surfactant is two-fold; it acts as a structure directing agent and it also allows for an

Figure 9.10 SEM and TEM images of porous silica made by using block copolymer and cellulose nitrate (a) and (b), respectively. (Reprinted with permission from: X. S. Li *et al.*, *Inorg. Chem. Commun.*, 2006, **9**, 7. Copyright 2006, Elsevier.)

efficient dispersal of CN in the solution of sol−gel precursors. Attempts to make mesoporous films without surfactant have failed. It also worth noting that the CN decomposition causes the formation of a harder skin on the top of the coating that appears less porous and, at least for specific compositions, crack-free.

Another *string to the nanotechnologist's bow* for controlling the pore architecture is microphase separation. This technique, combined with EISA, has shown high potential in terms of versatility and tuning. In fact, microphase separation can be applied to a large variety of systems and materials and it can produce very different results depending on the processing parameters. This is of particular importance for thin film deposition, a technique that is highly sensitive to environmental conditions such as relative humidity and temperature. In addition, thin film preparation offers the possibility to 'freeze' the system into a specific metastable state by applying a thermal shock or a fast solvent evaporation.

Another approach is based on the use of oil-in-water emulsions to form macropores into semi-crystalline mesoporous titania films.[31] The synthesis of mesoporous titania in aqueous media employs titanium tetraisopropoxide and Pluronic triblock copolymers with the addition of 1,3,5-triisopropylbenzene to induce the emulsion formation. The hierarchical titania porous structure obtained by this procedure shows three types of pores: evaporation-induced mesopores, emulsion-induced macropores and 'pinhole-like' macro-craters. The latter seem to play a key role in retaining the mesopore organization after titania crystallization. Calcined hierarchical structures maintain the mesoporous organization because the expansion of pinhole-like pores during thermal treatment counterbalances the tensile stress induced by shrinkage. However, mesoporous titania prepared under similar conditions is not stable to thermal treatment and loses long range mesopore organization.

Another synthesis of hierarchical titania films produced by microphase separation[32] uses as precursors titanium tetrachloride, Pluronic F127 block copolymer and low weight poly(propylene glycol) (PPG) dissolved in a mixture of butanol (BuOH) and tetrahydrofuran (THF). The composition of the mixture is of particular importance because it allows for a fine control of the final porous structure. PPG has a key role because it has a similar molecular weight to the Pluronic F127 polypropylene oxide block, BuOH is a good solvent for PPG while THF allows for increasing the film drying rate during deposition. By using these reagents, a tailored templating system with mono-, bi- or tri-modal pore distribution can be induced in the titania film. In fact, without adding THF to the porous titania solution of precursors, the dried films show bimodal porous structure as a result of self-assembly and phase separation. The pores do not show a high degree of organization and have an average size of 15 and 38 nm. The addition of THF to the mixture favors the dissolution of PPG because of the intermediate hydrophilicity of this solvent, and the system forms a uniform population of intermediate size pores with average diameter of 20 nm. In some particular cases, a third population of pores can be observed, producing a complex 'flower-like' pattern such as those shown in the **Figure 9.11**. The results have been subsequently confimed for similar systems.[33]

Another approach to the formation of porous silica films with hierarchical structure by controlling *in situ* two self-assembly processes is the formation of sodium chloride salt nanocrystals during self-assembly of a mesoporous hybrid-silica film.[34] After pore opening, the films show mesopores that appear to be well organized in a body-centered cubic fashion with size of 6 nm, and macropores with defined cubic shape in a range between 100 and 280 nm (**Figure 9.12**). The larger pores, the so-called nanoboxes, are not oriented but homogenously dispersed within the inorganic matrix. An easy way to avoid salt precipitation during film drying is to add disodium phosphate to the precursor solution, which induces the diffuse nucleation of nanoboxes during film formation and avoids the growth of NaCl crystals of large dimensions.[35] The process also offers the advantage of selective pore opening and therefore enables selective pore functionalization. The nanoboxes can be emptied by

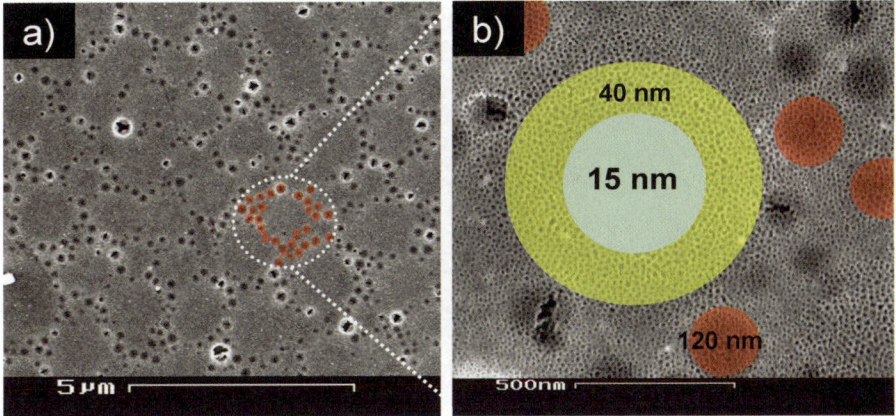

Figure 9.11   SEM images, at different magnifications, of titania films with tri-modal porosity. (Adapted with permission from L. Malfatti *et al.*, *Chem. Mater.*, 2009, **21**, 2763. Copyright 2009, American Chemical Society.)

washing the film with fresh water without removing the mesopore templates. However, the surfactant responsible for the mesopore formation can be degraded and removed by thermal treatment, without affecting the NaCl nanocrystals.

This is not all and aside from mesopores and nanoboxes, there are still other special features that can increase the complexity. In the previous material not only precipitation of crystalline salts can occur but also phase separation of the solvent with formation of hollow spheres as shown in **Figure 9.13**. The concurrency of the phenomena

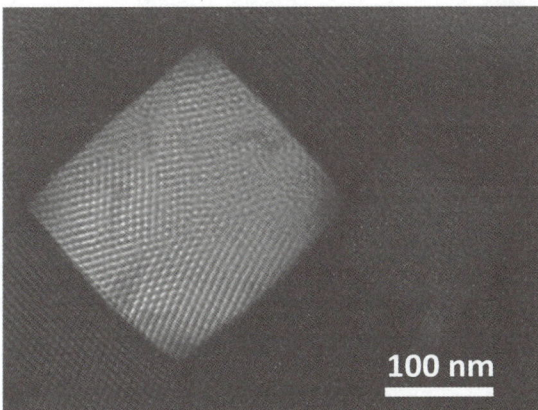

Figure 9.12   TEM images of bimodal porous silica displaying mesopores and 'nanocubes'. (Adapted with permission from L. Malfatti *et al.*, *Chem. Mater.*, 2009, **21**, 4846. Copyright 2009, American Chemical Society.)

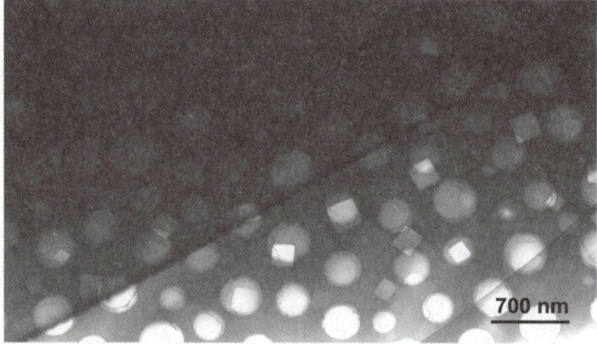

Figure 9.13   Nanoboxes and nanospheres in a hierarchical porous hybrid organic–inorganic silica film. (Adapted from ref. 36 with permission from the Centre National de la Recherche Scientifique (CNRS) and The Royal Society of Chemistry.)

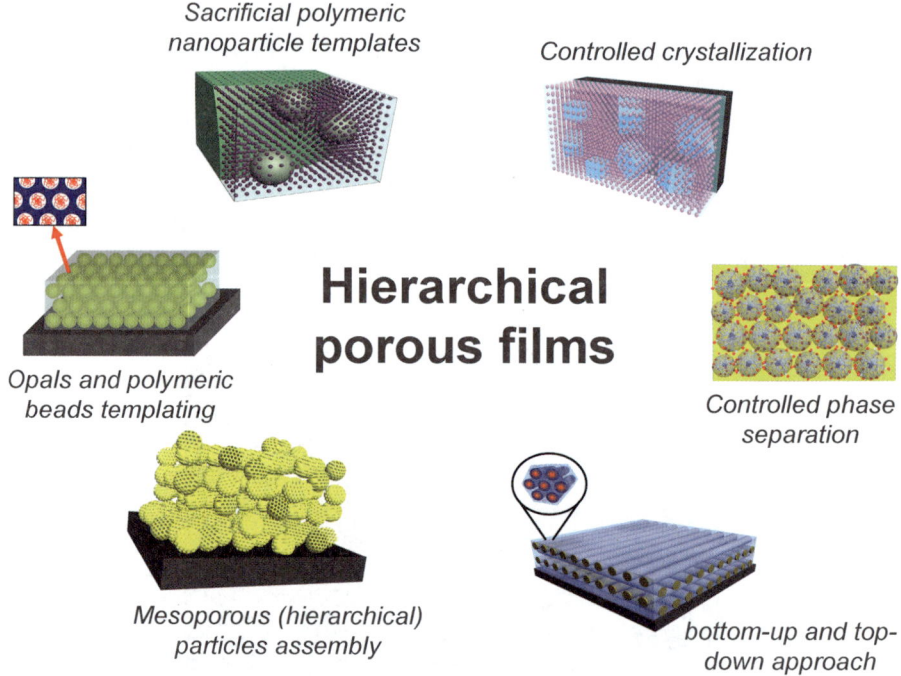

Figure 9.14   Scheme of the main routes that have been used so far to obtain hierarchical porous films by combining self-assembly with other templating strategies. (Adapted with permission from P. Innocenzi *et al.*, *Chem. Mater.*, 2011, **23**, 2501. Copyright 2011, American Chemical Society.)

gives a hierarchical material with pores of different shapes and dimensions. The mesopores are obtained *via* evaporation templated self-assembly, the cubes by precipitation of salts and the spheres *via* phase separation.[36] Only a specific combination of the synthesis parameters gives the three-modal distribution of porosities.

9.7   A HIERARCHICAL OVERVIEW

We have described so many different ideas and strategies for obtaining hierarchical porous structures in films. They can be divided into two main groups; one-pot synthesis routes and synthesis with pre-formed templating objects. A scheme of the main routes that have been proposed so far for obtaining hierarchical porous films by combining self-assembly with other templating strategies is shown in **Figure 9.14**. Multiple templating allows for obtaining pores of different dimensions, while combining all these strategies with self-assembly through evaporation is one of the most elegant approaches for obtaining hierarchical porous structures. Hierarchy is a concept that is strictly linked with complexity and is expected to increase *via* self-assembly. However, evaporation is very much a typical phenomenon of non-equilibrium systems and the combination of self-assembly with other processes is always full of surprises.

REFERENCES

1. Concise Oxford English Dictionary, Oxford University Press, UK, 2008.
2. S. J. Eichhorn, *Soft Matter*, 2011, **7**, 303.
3. D. Fengel and G. Wegener, *Wood Chemistry, Ultrastructure, Reactions*, Verlag Kessel, Remagen, 2003.
4. (a) O. Sel, D. Kuang, M. Thommes and B. Smarsly, *Langmuir*, 2006, **22**, 2311; (b) O. Sel, A. Brandt, D. Wallacher, M. Thommes and B. Smarsly, *Langmuir*, 2007, **23**, 4724.
5. P. Innocenzi, L. Malfatti and G. J. J. A. Soler-Illia, *Chem. Mater.*, 2011, **23**, 2501.
6. R. A. Caruso and M. Antonietti Adv. *Funct. Mater.*, 2002, **12**, 307.
7. (a) F. Heinroth, S. Münzer, A. Feldhoff, S. Passinger, W. Cheng, R. Carsten, B. Chichkov and P. Behrens *J. Mater. Sci.*, 2009, **44**, 6490; (b) F. Heinroth, I. Bremer, S. Münzer, P. Behrens, C. Reinhardt, S. Passinger, C. Ohrt and B. Chichkov, *Microporous Mesoporous Mater.*, 2009, **119**, 104.
8. The term is somehow misleading in this context. Microfabrication, in fact, is a word which comes from the field of lithography and refers to material shaping at the micron-scale. However, by following the IUPAC classification, *micron*-scale pores should be called *macro*-pores. This leads

to the apparently contradictory sentence 'micro-fabrication of the materials allows for macropore formation'.

9. P. C. Angelomé, M. C. Fuertes and G. J. A. A. Soler-Illia, *Adv. Mater.*, 2006, **18**, 2397.
10. M. C. Fuertes and G. J. A. A. Soler-Illia, *Chem. Mater.*, 2006, **18**, 2109.
11. A. Zelcer, A. Wolosiuk and G. J. A. A. Soler-Illia, *J. Mater. Chem.*, 2009, **19**, 4191.
12. X. Du and J. He, *Langmuir*, 2010, **26**, 13528.
13. X. Du and J. He, *J. Colloid Interface Sci.*, 2010, **345**, 269.
14. Lotus leaves are water repellent because their surface is made up of micron-sized knobbles, which are covered by nanostructures with waxy tips. This explains why the hydrophobicity induced by surface roughness is called the lotus effect.
15. X. Du, X. Li and J. He, *ACS Appl. Mater. Interfaces*, 2010, **2**, 2365.
16. F. Li and A. Stein, *Chem. Mater.*, 2010, **22**, 3790.
17. P. Falcaro, L. Malfatti, T. Kidchob, G. Giannini, A. Falqui, M. F. Casula, H. Amenitsch, B. Marmiroli, G. Grenci and P. Innocenzi, *Chem. Mater.*, 2009, **21**, 2055.
18. B. O'Regan and M. Grätzel, *Nature*, 1991, **353**, 737.
19. G. Kaune, M. Memesa, R. Meier, M. A. Ruderer, A. Diethert, S. V. Roth, M. D' Acunzi, J. S. Gutmann and P. Müller-Buschbaum, *ACS Appl. Mater. Interfaces*, 2009, **1**, 2862.
20. J. S. Gutmann, P. Müller-Buschbaum and M. Stamm, *Faraday Discuss.*, 1999, **112**, 285.
21. T. Kimura, N. Miyamoto, X. Meng, T. Ohji and K. Kato, *Chem. – Asian J.*, 2009, **4**, 1486.
22. O. Sel, S. Sallard, T. Brezesinski, J. Rathouský, D. R. Dunphy, A. Collord and B. Smarsly, *Adv. Funct. Mater.*, 2007, **17**, 3241.
23. M. Etienne, S. Sallard, M. Schroder, Y. Guillemin, S. Mascotto, B. M. Smarsly and A. Walcarius, *Chem. Mater.*, 2010, **22**, 3426.
24. A. Walcarius, E. Sibottier, M. Etienne and J. Ghanbaja, *Nat. Mater.*, 2007, **6**, 602.
25. S. Tao, J. Yin and G. Li, *J. Mater. Chem.*, 2008, **18**, 4872.
26. M. Sakurai, A. Shimojima, M. Heishi and K. Kuroda, *Langmuir*, 2007, **23**, 10788.
27. P. Yang, T. Deng, D. Zhao, P. Feng, D. Pine, B. F. Chmelka, G. M. Whitesides and G. D. Stucky, *Science*, 1998, **282**, 2244.
28. E. Kim, Y. Xia and G. M. Whitesides, *Adv. Mater.*, 1996, **245**, 8.
29. (a) X. S. Li, G. E. Fryxell, C. Wang and J. Young, *Inorg. Chem. Commun.*, 2006, **9**, 7; (b) R. E. Williford, G. E. Fryxell, X. S. Li and R. S. Addleman, *Microporous Mesoporous Mater.*, 2005, **84**, 201; (c) X. S. Li, G. E. Fryxell, C. Wang and J. Young, *Microporous Mesoporous Mater.*, 2007, **99**, 308.
30. W. S. Dutton, *DuPont: One Hundred and Forty Years*, Charles Scribner and Sons, New York, 1942, pp. 153–169.

31. X. Meng, T. Kimura, T. Ohji and K. Kato, *J. Mater. Chem.*, 2009, **19**, 1894.
32. L. Malfatti, M. G. Bellino, P. Innocenzi and G. J. J. A. Soler-Illia, *Chem. Mater.*, 2009, **21**, 2763.
33. Q. L. Wu, N. Subramanian and S. E. Rankin, *Langmuir*, 2011, **27**, 9557.
34. L. Malfatti, P. Falcaro, D. Marongiu, M. F. Casula, H. Amenitsch and P. Innocenzi, *Chem. Mater.*, 2009, **21**, 4846.
35. L. Malfatti, D. Marongiu, H. Amenitsch and P. Innocenzi, *J. Phys. Chem. C*, 2011, **115**, 12702.
36. P. Innocenzi L. Malfatti, D. Marongiu and M. F. Casula, *New J. Chem.*, 2011, **35**, 1624.

Subject Index